# Value Engineering Synergies with Lean Six Sigma

## Combining Methodologies for Enhanced Results

# Value Engineering Synergies with Lean Six Sigma

## Combining Methodologies for Enhanced Results

Jay Mandelbaum ◆ Anthony Hermes
Donald Parker ◆ Heather Williams

CRC Press is an imprint of the
Taylor & Francis Group, an **informa** business
A PRODUCTIVITY PRESS BOOK

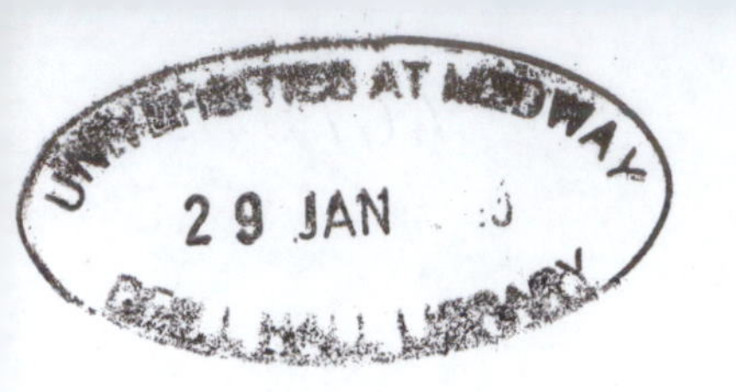

CRC Press
Taylor & Francis Group
6000 Broken Sound Parkway NW, Suite 300
Boca Raton, FL 33487-2742



Printed in the United States of America on acid-free paper
Version Date: 20111208

International Standard Book Number: 978-1-4665-0201-7 (Paperback)



**Library of Congress Cataloging-in-Publication Data**

Value engineering synergies with lean six sigma : combining methodologies for enhanced results / Jay Mandelbaum ... [et al.]. p. cm.
Includes bibliographical references and index.
ISBN 978-1-4665-0201-7 (pbk. : alk. paper)
1. Production management. 2. Industrial management. 3. Value analysis (Cost control) 4. Six sigma (Quality control standard) I. Mandelbaum, Jay.

TS155.V194 2012
658.4'013--dc23 2011048718

**Visit the Taylor & Francis Web site at**
**http://www.taylorandfrancis.com**

**and the CRC Press Web site at**
**http://www.crcpress.com**

# Contents

# Preface

It's difficult to identify the exact point in time when the idea for this book actually began to form. That's primarily because everything was subliminal—there was no conscious plan to move in that direction.

Using imperfect hindsight, the earliest underlying driving factor for writing this book occurred around 2003 as a result of observing the experiences of Value Engineering (VE) practitioners trying to combine its use with Lean and Six Sigma. The unambiguous impression from these observations was that business improvement initiatives compete with one another. Organizations are often built around one approach and they are not always tolerant of other approaches. The principal reasons for this are

- the initiative du jour usually has greater management support;
- career advancement sometimes depends upon getting ahead of others; and
- business plans often seek competitive advantage by emphasizing their "unique" approach that differentiates them from alternatives.

This initial impression was reinforced a few years later as a result of research in the area of quality management. Stovepipes were found to exist in companies among continuous process improvement, Lean Six Sigma (LSS), and quality organizations. Perhaps the "last straw" in this progression occurred in the formation of a Continuous Process Improvement organization in the Office of the Secretary of Defense (OSD). The Department of Defense (DOD) VE lead did not attempt to include VE in the charter of that new office.

In the 2008–2009 timeframe, an LSS master black belt became the DOD VE lead. He asked the question about the relationship between VE and LSS. The answer given was not entirely adequate. The question was perceived in part as a threat, so the response in part reflected that perception—too much of an emphasis on complementarity and not enough emphasis on synergy.

The overt idea for writing about the synergies actually emerged when an LSS process improvement person was transferred into the DOD office that had VE responsibilities, although the actual beginning of the writing process occurred much later. In 2010, a paper was completed for OSD on the synergies. The contents of that paper comprise the bulk of this book. But that's not when the inspiration for the book emerged.

The VE methodology was originally formulated during World War II when the General Electric (GE) Company was looking for substitutes for materials and labor that were in short supply. GE discovered that alternative approaches were often less costly and performed better. Lawrence Delos Miles is credited with the development of the methodology.

One of the peer reviewers of the synergy paper was a GE Aviation engineer who had some of Miles' original training material. He was also familiar with LSS and had been a part of the adoption of Design for Six Sigma at GE. He was the one who suggested publishing the work as a book because of the limited material that had been written on the subject.

The authors wish to thank all of the reviewers of that 2010 paper—Karen Richter and Lance Roark, research staff members from the Institute of Defense Analyses; S. Jeff Hestop, a master black belt from General Electric Aviation; and Gene Wiggs, a consulting engineer from General Electric Aviation, who was both a reviewer and a contributor.

But that's not the end of the story. When the synergy arguments were presented to OSD's Strategic Management and Performance Office (the new name of the former Continuous Process Improvement Office), they were immediately embraced as an aspect of LSS black belt training. As a result, the training material provided in Appendices C, D, and E were also developed.

The authors' fervent hope is that government and industry will embrace the ideas and synergies presented in this book to accelerate efficiencies in their operations.

# Abbreviations

**ANOVA:** analysis of variance
**AoA:** Analysis of Alternatives
**AT&L:** Acquisition, Technology, and Logistics
**CAM-I:** Consortium for Advanced Management International
**CIWS:** Close-in Weapon System
**CJCSI:** Chairman of the Joint Chiefs of Staff Instruction
**CLIN:** contract line item
**CTQ:** critical to quality
**DFSS:** Design for Six Sigma
**DMADV:** Define, Measure, Analyze, Design (and Optimize), and Verify
**DMAIC:** Define, Measure, Analyze, Improve, and Control
**DMSMS:** diminishing manufacturing sources and material shortages
**DOD:** Department of Defense
**DODI:** Department of Defense Instruction
**DOE:** design of experiments
**DOTMLPF:** doctrine, organization, training, materiel, leadership and education, personnel, or facilities
**DSCC:** Defense Supply Center Columbus
**FAST:** Function Analysis System Technique
**FMEA:** failure modes and effects analysis
**GE:** General Electric
**GRR:** gage repeatability and reproducibility
**HVAC:** heating, ventilation, and air conditioning
**ICAM:** Improved Chemical Agent Monitor
**IDOV:** identify, design, optimize, and verify
**JIT:** just-in-time
**LSS:** Lean Six Sigma
**MIT:** Massachusetts Institute of Technology
**OMB:** Office of Management and Budget

**QFD:** Quality Function Deployment
**RACI:** Responsible, Approval, Contributor, and Informed
**ROI:** return on investment
**SIPOC:** Supplies, Input, Process, Outputs, and Customers
**SMART:** specific, measurable, achievable, relevant, and time-bound
**SWOT:** strengths, weaknesses, opportunities, and threats
**TOC:** Theory of Constraints
**TQM:** Total Quality Management
**TRIZ:** Theory of Inventive Problem Solving (in Russian, *Teoriya Resheniya Izobretatelskikh Zadatch*)
**VA:** Value Analysis
**VE:** Value Engineering
**VECP:** Value Engineering Change Proposal
**VEP:** Value Engineering Proposal
**VM:** Value Management
**VSM:** value stream mapping

# Executive Summary

Lean Six Sigma (LSS), its Design for Six Sigma (DFSS) variant, and Value Engineering (VE) were developed as business process improvement initiatives. This book explores synergies among LSS, DFSS, and VE by identifying opportunities where they can be used together to increase the likelihood of obtaining improvements beyond the capability of just one approach.

The origins of these initiatives are different. VE originated in the industrial community during World War II when many manufacturers were forced to substitute materials and designs as a result of critical material shortages. LSS is a combination of Lean, Six Sigma, and the Theory of Constraints (TOC). Each of these components also has different origins. Lean concepts can be traced to the evolution of the Toyota Production System in the decades following World War II. Six Sigma has its genesis in the application of probability theory to statistical quality control. TOC represents a paradigm shift to improve the concepts of just-in-time (JIT) and Total Quality Management (TQM) to help stimulate the needed change. DFSS was developed to apply Six Sigma principles in the design phase.

These differences in origin lead to varying approaches to problem solving. Each initiative has different phases in its methodological approach:

- VE phases are orientation, information, function analysis, creative, evaluation, development, presentation, and implementation.
- LSS phases are define, measure, analyze, improve, and control.
- DFSS phases are define, measure, analyze, design (and optimize), and verify.

Business process improvement initiatives are also cyclical in nature. They evolve over time and can ultimately be replaced by processes that attempt to integrate specific attributes of older initiatives with the latest approaches and/or technologically enabled methodologies. Practitioners often differentiate their initiatives from others because of different origins, vocabulary, skills, and training; effectiveness in particular circumstances; and applicability to a specific problem. Unfortunately, these differentiations are not always important and can create organizational stovepipes that compete with one another. A successfully implemented methodology may not be the best and only one for every problem. Depending on the situation, integrating multiple approaches can provide valuable ideas and insights that augment the benefits of using the approaches separately. Such synergies not only achieve better results, but also break down the organizational stovepipes that naturally occur when different offices are assigned responsibility for different problem-solving methods.

To examine these synergies, this book describes the steps and activities within each of the methodological phases to provide the reader an appreciation for the logical flow of events that transition smoothly from one activity to another, working toward a solution. These descriptions are also used to identify similarities and differences and construct a cross-reference mapping between VE and LSS. The differences do not imply that one methodology is better than the other, nor do they imply weaknesses. Instead, the differences indicate opportunities where both approaches may be used together to achieve better results.

## How VE Can Benefit from LSS/DFSS

When LSS establishes goals, customer communication tools such as Likert scales, surveys, interviews, and focus groups are used. The VE counterpart, prioritize issues, is more focused on potential gains and feasibility of implementation. More formalized customer communication would help with decision-maker acceptance and approval of VE-generated recommendations.

LSS has a more detailed front-end process for data collection. Whereas the VE methodology simply states that the data should be collected, LSS creates and analyzes process maps, determines and prioritizes measurement systems, and establishes a formal data-collection plan. When VE finalizes the problem and facts, it often uses a Quality Function Deployment (QFD) tool to obtain a better understanding of the data and data sources in the context of the problem. The LSS's Supplies, Inputs, Process, Outputs, and Customers (SIPOC) framework is used to understand the entire process and where the problem fits in. VE's use of SIPOC could add insight to its function analysis process.

LSS also has a more disciplined approach toward implementation. VE simply creates an implementation plan and follows typical best practices to execute it. The LSS control plan is a formal activity designed to ensure that execution proceeds as planned and with specific metrics identified in advance. Furthermore, LSS includes a formal corrective action plan (sometimes as a separate process), which is not an unambiguous part of the VE methodology.

These differences represent areas where incorporating some LSS features would likely improve the VE methodology. These synergies would help formalize the VE process to reduce the likelihood of overlooking important information needed to help determine a course of action. They would also improve the likelihood of successful implementation.

## How LSS/DFSS Can Benefit from VE

VE and LSS develop solutions to problems from different perspectives. Some of the most important distinctions are as follows:

- VE explicitly considers cost by collecting cost data and using cost models to make estimates for all functions over the life cycle. LSS reduces cost by eliminating waste and reducing variation through the use of statistical tools on process performance data. Exclusive emphasis on waste can be contradictory to reducing life-cycle cost. In VE, some waste can be tolerated if it is necessary to achieve a function that reduces the life-cycle cost. Safety stock to mitigate occasional supply disruption is a good example.
- In determining what should be changed, VE's function analysis identifies areas that cost more than they are worth, while LSS identifies root causes of problems or variations. VE's separation of function from implementation forces engineers to understand and deliver the requirements.
- For required functions that cost more than they are worth, VE uses structured brainstorming to determine alternative ways of performing them. LSS brainstorms to identify how to fix the root causes. Because functional thinking is not the common way of examining products or processes, VE augments the structured innovation process in a way that generates a large number of ideas. Enormous improvements are possible by determining which functions are really required and then determining how to best achieve them.

- VE develops solutions by evaluating the feasibility and effectiveness of the alternatives. LSS emphasizes solutions that eliminate waste and variation and sustain the achieved gains. VE eliminates waste in a different way. VE separates the costs required for basic function performance from those incurred for secondary functions to eliminate as many non-value-added secondary functions as possible, improve the value of the remaining ones, and still meet the customer requirements.
- An LSS focus on quick wins may preclude an in-depth analysis of the situation. Without analysis, projects can suboptimize or even work in opposition to one another. Using function analysis should prevent this suboptimization.

While DFSS is a proactive and anticipatory approach that helps evaluate and optimize conceptual, preliminary, and detailed designs, it is not an automatic process and does not replace skilled designers. Developing an effective design that does everything a user wants from a performance perspective and from the perspective of design considerations (e.g., supportability, maintainability, information assurance, availability, reliability, producibility may be applicable), while not costing too much or weighing too much, will almost always benefit from the group perspectives and discussions of the Function Analysis and Creative phases of the VE job plan. VE links the customer requirements to the design to manage cost.

The literature on LSS and VE compares the strengths and weaknesses of the methodologies and highlights opportunities for collaboration. The literature examining these methodologies points to two primary areas where VE can contribute: scope and creative tools such as the Function Analysis System Technique (FAST) diagram. Experts are encouraging about the prospects for synergizing the methodologies, particularly in a process where a team can take advantage of respective strengths and avoid respective weaknesses.

The highest leverage points for VE contributions to LSS and DFSS over a life cycle vary by application. For a product, VE can provide benefit everywhere—from concept to decision to operations and support. For a service, VE is most applicable during conceptual design and operations. For a construction project, primary VE opportunities occur during preliminary and detailed design.

## Recommendations

Both LSS and VE have unique attributes and perspectives for process improvement. Since certain problems may be more readily, effectively, or thoroughly managed by using one or both of these perspectives, exploring the full range of solution options is crucial. A comparison of the methodological approaches and the examples of synergies discussed in the literature leads to the conclusion that VE techniques are sometimes better equipped to lead to improvements or solutions complementary to those identified through a Define, Measure, Analyze, Improve, and Control (DMAIC)/DFSS approach. These opportunities for synergy include the following:

- **Function Analysis and the FAST diagram.** The disciplined use of function analysis is the principal feature that distinguishes the value methodology from other improvement methods. Function analysis challenges requirements by questioning the existing system and critical thinking. Function analysis subsequently develops innovative solutions to revised requirements.
- **Cost Focus.** VE only develops alternatives that provide the necessary functions. By examining only those functions that cost more than they are worth and identifying the total cost of each alternative, VE explicitly lowers cost and increases value.

VE does not take the place of LSS efforts, but it does present significant opportunities to enhance LSS-developed options. Therefore, LSS training can be augmented to include the VE approach to function analysis, creativity, and associated elements of evaluation and development to identify candidate solutions as part of the Analyze and Improve phases of DMAIC.

As far as DFSS is concerned, VE tools should be explicitly used in the process. They should be used in the Analyze phase of Define, Measure, Analyze, Design (and optimize), and Verify (DMADV) to construct function views of the product or process to identify customer priorities and determine functional requirements. They should also be used in the Design phase of DMADV to generate alternative design concepts and to modify component/subsystem preliminary and detailed designs to introduce new elements to the evaluation and optimization processes.

# Chapter 1

# Introduction

Lean Six Sigma (LSS), its Design for Six Sigma (DFSS) variant, and Value Engineering (VE) were developed as business process improvement initiatives. This book explores synergies among LSS, DFSS, and VE by identifying opportunities where they can be used together to increase the likelihood of obtaining improvements beyond the capability of just one approach.

The origins of these initiatives are different. VE originated in the industrial community during World War II when critical material shortages forced many manufacturers to substitute materials and designs. When the General Electric (GE) Company found that many of the substitutes were providing equal or better performance at less cost, it launched an effort in 1947 to improve product efficiency by intentionally and systematically developing less costly alternatives. Lawrence D. Miles, a staff engineer for GE, led this effort. Miles combined several ideas and techniques to develop a successful methodological approach for ensuring value in a product. The concept quickly spread through private industry as the possibilities for large returns from relatively modest investments were recognized. This methodology was originally termed *Value Analysis (VA)* or *Value Control*.

LSS is a combination of Lean, Six Sigma, and the Theory of Constraints (TOC). Each of these components also has different origins.

Lean concepts can be traced to the evolution of the Toyota Production System in the decades following World War II.* They became established in the Western world in the 1980s and 1990s. "Lean thinking is the dynamic, knowledge-driven, and customer-focused process by which all people in a

* Refer, for example, to James P. Womack, Daniel T. Jones, and Daniel Roos, *The Machine That Changed the World* (New York: Rawson Associates, 1990).

defined enterprise continuously eliminate waste with the goal of creating value."[*] Value creation is a central concept in Lean thinking to build robust, adaptive, flexible, and responsive enterprises.

Six Sigma has its genesis in the application of probability theory to statistical quality control. The goal of Motorola's Six Sigma initiative was to identify and reduce all sources of product variation—machines, materials, methods, measurement systems, the environment, and the people—in the process. The idea is not new. It can be traced to the introduction of Lean thinking and Total Quality Management (TQM). At a technical level, Six Sigma is aimed at achieving virtually defect-free operations, where parts or components can be built to very exacting performance specifications. Underlying Six Sigma as a statistical concept[†] is the construct of standard deviation, a measure of dispersions around the mean. Reducing variation to the Six Sigma level denotes reaching a performance level of 99.99966% perfection (3.4 defects or nonconformance per million opportunities[‡]). This level of performance means virtually defect-free production, where a defect is defined as any instance or event in which the product fails to meet a customer requirement.

TOC was developed by Eliyahu M. Goldratt, a physicist by education, in a series of publications over the past two decades.[§] According to Goldratt, TOC represents a paradigm shift to improve the concepts of just-in-time (JIT) and TQM to help stimulate the needed change. The important contribution of TOC has been its recognition at a conceptual level that systems should be viewed as *chains* of interdependence and that systems contain leverage points—constraints—where proactive change initiatives can deliver large positive effects on overall system performance.

DFSS was developed to apply Six Sigma principles in product design. A common rule of thumb is that only 20% of cost can be affected by improving the efficiency of processes, while 80% of costs are locked in during design. Consequently, improving the design early in the life cycle, when the

---

[*] Earl M. Murman et al., *Lean Enterprise Value: Insights from MIT's Lean Aerospace Initiative* (Houndmills, Basingstoke, Hampshire RG21 6XS, Great Britain: Palgrave, 2002), 90.

[†] Industry has a long history of using statistics. See, for example, Gerald J. Hahn, *The Role of Statistics in Business and Industry* (Hoboken, NJ: John Wiley and Sons, 2008).

[‡] Defects per million opportunities indicates how many defects would be observed if an activity were repeated a million times.

[§] See, for example, Eliyahu M. Goldratt, *Theory of Constraints* (Croton-on-Hudson, NY: North River Press, Inc., 1990).

design flexibility is highest, has far greater leverage.* Historically, DFSS was created in part because Six Sigma organizations found that they could not optimize products (or their manufacturing process) past three or four sigma without fundamentally redesigning the product. This means that *Six Sigma* levels of performance have to be *built-in* or *by design*. While Six Sigma requires a process to be in place and functioning, the objective of DFSS is to determine the needs of the customers and the business and to drive those needs into the product/process solution. It is product/process *generation* as opposed to *improvement*. DFSS aims to create a product/process by optimally building the efficiencies of Six Sigma methodology into the product/process before implementation.

Business process improvement initiatives are also cyclical in nature. They evolve over time and can ultimately be replaced by processes that attempt to integrate specific attributes of older initiatives with the latest approaches and technologically enabled methodologies. Practitioners often differentiate their initiatives from others because of different origins, vocabulary, skills, and training; effectiveness in particular circumstances; and applicability to a specific problem. Unfortunately, these differentiations are not always important, and can create organizational stovepipes that compete with one another. A successfully implemented methodology may not be the best and only one for every problem. Depending on the situation, integrating multiple approaches can provide valuable ideas and insights that augment the benefits of using the approaches separately. Such synergies not only achieve better results, but also break down the organizational stovepipes that naturally occur when different offices are assigned responsibility for different problem-solving methods.

To examine these synergies, this book is organized as follows:

- Chapter 2 discusses the VE methodology.
- Chapter 3 discusses the LSS and DFSS approaches.
- Chapter 4 cross-references the methodologies and identifies ways in which one methodology can benefit the other.
- Chapter 5 examines opportunities for synergy in more detail.
- Chapter 6 contains some final remarks.
- Appendix A contains detailed LSS-VE cross-reference charts

* For example, Hinckley states that the cost of change is 100 times higher during production tooling than during conceptual design in C. Martin Hinckley, *Managing Product Complexity: It's Just a Matter of Time*, Report No. SAND-98-8564C (Livermore, CA: Sandia National Laboratories, June 1, 1998).

- Appendix B lists some common LSS, DFSS, and VE tools.
- Appendix C is general VE material designed for LSS black belt training.
- Appendix D is product-oriented VE material designed for LSS black belt training.
- Appendix E is process-oriented VE material designed for LSS black belt training.

# Chapter 2

# The Value Engineering (VE) Methodology

The VE methodology, also referred to as the job plan, is divided into eight phases:

- Orientation
- Information
- Function Analysis
- Creative
- Evaluation
- Development
- Presentation
- Implementation

The following sections describe each phase and its purpose. Figure 2.1 graphically depicts the phases and the principal steps within the job plan. The application of the methodology to a problem is often referred to as a value study. Except for the Orientation and Implementation phases, the value study typically occurs in a workshop setting.

## A. Orientation Phase

The Orientation phase refines the problem statement and prepares for the workshop. The value study and workshop have a greater likelihood

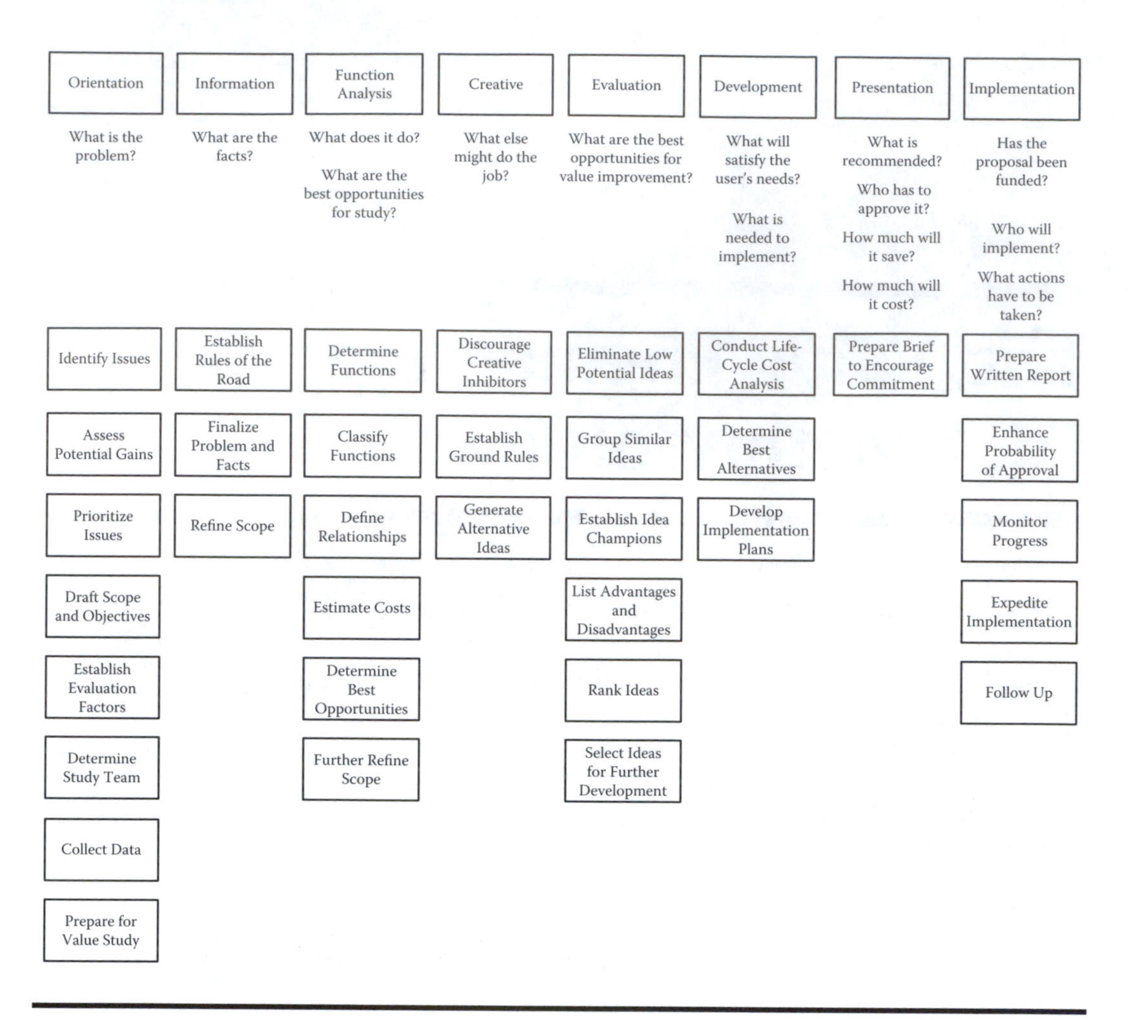


**Figure 2.1 VE job plan.**

of success if ample preparation time has been devoted to determining what aspects of the problem will be addressed in detail and preparing everything needed for the analysis. Throughout these preparatory activities, a close working relationship between the study team leader and the manager sponsoring the VE project contributes significantly to a successful outcome.

The following subsections describe the activities during the Orientation phase. The activities can occur in an order different from that shown here. Some activities can also be repeated or occur simultaneously if other people are supporting the team leader's efforts.

The first five activities represent one systematic approach to refining the problem statement. The job plan can also be used entirely in the context of the Orientation phase as a formal project planning tool.

## 1. Identify Specific Issues to Be Addressed

The first step in a project is to identify a problem. The problem area should be divided into its constituent elements. Each element should represent a specific issue that can be addressed and resolved.

Consider, for example, the Navy's Standard Missile program. Missile demand was level, but the price was increasing while budgets were decreasing. Of the three controllable constituent elements of missile cost (production, development, and logistics), production costs were determined by the program office to be the most fruitful area for further investigation. In fact, the production costs could readily be broken down into smaller and smaller constituent elements to form the basis of individual VE projects.*

Identifying specific issues is accomplished by developing an understanding of the sponsor's problems and avoiding areas that the sponsor would not be able to change because of political, cultural, or feasibility implications. Once the problems are understood, they can be addressed at varying levels of detail. At this stage of the VE methodology, an adequate amount of detail is needed to obtain a general grasp of potential VE projects for the issue under consideration.

## 2. Assess Potential Gains for Resolving Each of These Issues

The purpose of this activity is to identify issues that have the greatest potential for value improvement. Solution areas postulated this early should be used only for this step because they could inhibit creative activities used later in the job plan to generate alternatives.

The assessment of the potential gains for resolving issues should be as quantitative as possible; however, at this stage of the analysis, estimates will be crude. While developing a reasonable understanding of the costs involved may not be too difficult, savings estimates are much more problematic since no solution has been developed. Some information is normally available, however, and should be used to assess the problems and potential gains.

In the Standard Missile example, one of the VE projects involved the transceiver assembly. One potential solution was to replace the assembly

* See Roland Blocksom, "STANDARD Missile Value Engineering (VE) Program—A Best Practices Role Model," *Defense AT&L Magazine*, July–August 2004, 41–45.

with a less costly one. Savings estimates were difficult to obtain because the characteristics of the new assembly were unknown. Another potential solution involved developing a greater level of aggregation. Here, savings would be generated by eliminating tests.

## 3. Prioritize Issues

While prioritization should weigh the potential gains, it should also consider the likelihood of determining an effective solution and the feasibility of implementing that solution. In the case of the transceiver assembly for the Standard Missile, the second potential solution (developing a greater level of aggregation) was much more straightforward and had a higher likelihood of success than the first potential solution (replacing the assembly with a less costly alternative).

Understanding the importance of the problem to the project sponsor is also a key factor. If the sponsor is determined to solve the problem, the likelihood of success is enhanced. Once management commitment is understood, a useful question to ask is why the problem was not already solved.

The answer to this question may identify roadblocks to overcome. Knowing what stands in the way of a solution is another important consideration for prioritization. Finally, other benefits, such as performance improvement, should be considered.

## 4. Draft a Scope and an Objective for the Value Study

The study team's efficiency is significantly enhanced when limits are established in advance. More than one of the constituent problem elements can be included in the scope. The study sponsor must approve the scope. Ultimately, the scope and the objective will be finalized in the Information phase. This preliminary work will expedite finalization.

## 5. Establish Evaluation Factors

Targets for improvement should be challenging, and evaluation factors must be measurable. These factors determine the relative importance of the ideas and the potential solutions generated by the team. The study sponsor must approve the improvement targets and the evaluation factors.*

* In manufacturing-oriented workshops, criteria are not usually selected until competing alternatives have been developed.

## 6. Determine Team Composition

Essential team member characteristics include technical or functional expertise, problem-solving and decision-making abilities, and interpersonal skills. Participants should be team players who are willing to share responsibilities and accountability while working together toward a common objective. The team should also be multidisciplinary and include all factions affected by the study to ensure that relevant stakeholders and experts are included. J. Jerry Kaufman suggests that because gathering all the information needed to make a *no-risk decision* is impossible, a multidisciplinary team should provide enough different perspectives to at least substantially reduce the risk of ignoring a pertinent viewpoint.*

The ideal team size is five to seven people. A team with more than ten participants is difficult to control.† After the team members have been selected, the team leader should prepare a management memorandum to be sent to all team members. This memorandum should do the following:

- Emphasize the importance of their role
- Approve the necessary time commitment
- Authorize sharing of any objective and subjective data that bear on the problem
- Identify the team leader

## 7. Collect Data

The team leader organizes the data-collection activities in advance of the workshop. As more information is brought to bear on the problem, the probability of substantial benefit increases. To increase the study team's productivity, collecting as much data as possible in advance is crucial. The data-collection effort benefits from having the entire team involved. In fact, some team members may have key information readily available.

The data should be as tangible and quantitative as possible and should include anything potentially useful for understanding the problem, developing solutions, and evaluating the pros and cons of the solutions. The paramount considerations are getting enough facts and getting them from reliable sources.

---

* J. Jerry Kaufman, *Value Engineering for the Practitioner* (Raleigh, NC: North Carolina State University, 1990), 2–4.

† If more participants are needed, the use of on-call experts should be considered.

In addition to possessing specific knowledge of the item or process under study, the team should have all available information concerning the technologies involved and should be aware of the latest technical developments pertinent to the subject being reviewed.

Developing and ranking alternative solutions depends on having reliable cost data. Data on customer and user attitudes also plays a key role. Part of the VE study seeks to identify which aspect of the task holds the greatest potential for payoff. This potential for payoff is a function of the importance to the user and customer. The seriousness of user-perceived faults is also a factor in prioritization.

## *8. Prepare Logistically for the Value Study*

The VE study facilitator, who may also be the team leader, prepares the team to participate in the study. He/she is normally certified by SAVE, the VE professional society. The two levels of certification are Certified Value Specialist and Associate Value Specialist.

Initially, brief meetings with potential team members can be held to determine who should participate. The team leader/facilitator should do the following:

- Ensure that participants know what data they should bring.
- Set up study facilities and prepare materials (easels, markers, and so forth).
- Set up a kickoff briefing and results briefing with management.
- Obtain an example of a study item for the team to use.

Prestudy reading materials should be identified and distributed to the participants. Materials that can be assigned as advanced reading include the agenda, operational requirements documents, design documents (drawings and specifications), performance requirements, production quantities, inventory data, failure/quality information, and other documents necessary to ensure consistent understanding of the issues.

A preworkshop orientation meeting might be useful to accomplish the following:

- Review workshop procedures.
- Acquaint the team with the problem and read-ahead material.
- Eliminate incorrect preconceived notions about VE, the job plan, the workshop, the problem, the people, and so forth.

- Jump-start the team-building process.
- Clarify acceptable and unacceptable behaviors (i.e., "rules of the road") for team-member participation.
- Identify additional information needs.

The date should be set reasonably far enough in advance (four to six weeks) to allow personnel to arrange their schedules around the study. When a workshop setting is used, the value study typically takes three to five days.*

## B. Information Phase

The Information phase finalizes the scope of the issues to be addressed, the targets for improvement, and the evaluation factors; collects and analyzes the data; and builds cohesion among team members. In many respects, the Information phase completes the activities begun in the Orientation phase. This work is normally conducted in a workshop setting and is often the first opportunity for all team members to come together (if no prework-shop orientation meeting was scheduled). Consequently, the Information phase should be used to motivate the team to work toward a common goal. Finalizing the scope of the issues to be addressed, the targets for improvement, the evaluation factors, and the data collection and analysis efforts are ideal endeavors for building team cohesion. The following subsections describe the activities during the Information phase.

### *1. Establish Workshop Rules of the Road*

This activity begins the team-building process; therefore, the facilitator should ensure that all team members know each other and their relevant backgrounds, authority, and expertise. Some authors suggest that team-building exercises should be conducted at the beginning of the workshop.† The following guidelines should be established to set the stage for an effective working relationship among the team members:

---

* Three days may be sufficient for small studies, but five days are more common. To avoid keeping team members away from their job for five consecutive days, a separate two-day workshop can be held for the Development and Presentation phases.

† Robert B. Stewart, *Fundamentals of Value Methodology* (Bloomington, IN: Xlibris Corporation, 2005), 113–118.

- Share workload equally whenever possible.
- Be willing to admit not knowing something, but strive to get the answer. Do not be afraid to make mistakes.
- Stay focused and follow the basic problem-solving steps. Do not waste time discussing whether to use each step; complete the steps and conduct an evaluation after completing the entire workshop. Be sure to understand the approach and its purpose, including the reason for each step and the technique being applied. Keep the discussions relevant.
- Work together as a team. Instead of forcing solutions—sell them! A problem can have multiple solutions.
- Be a good listener; do not interrupt or criticize others for what they say.
- Keep an open mind and do not be a roadblock.
- Be enthusiastic about the project and what it is that you are doing.
- Do not attempt to take over as a team leader; be as helpful as possible. The leader already has a difficult job in guiding, controlling, and coordinating the overall effort.
- Accept conflicts as necessary and desirable. Do not suppress or ignore them. Work through them openly as a team.
- Respect individual differences. Do not push each other to conform to central ideas or ways of thinking.
- Work hard. Keep the team climate free, open, and supportive.
- Fully use individual and team abilities, knowledge, and experience.
- Accept and give advice, counsel, and support to each other while recognizing individual accountability and specialization.

## 2. Finalize the Problem and the Associated Facts

Before starting the analysis, the team should finalize the problem statement to ensure mutual understanding. This process involves discussing the problem so that all team members achieve a consistent understanding of the issues. The focus should be on the specifics, not generalities. This approach also serves as a useful team-building exercise.

The VE team should begin collecting information before the start of the workshop. If possible, this information should include physical objects (e.g., parts) that demonstrate the problem. When supported facts cannot be obtained, the opinions of knowledgeable people can be used. These people can be invited to participate in the workshop, or their opinions can be documented. The Information phase is typically used to familiarize the team members with the data and the data sources in the context of defining the problem. The keys are as follows:

- Getting up-to-date facts from the best sources
- Separating facts from opinion
- Questioning assumptions

Having all of the pertinent information creates an ideal situation, but missing information should not preclude the performance of the VE effort.

Quality Function Deployment (QFD) is a structured approach to translating customer needs or requirements into specific plans to produce products or develop processes to meet those needs.* Henry A. Ball suggests that QFD techniques can be beneficial in the Information phase because a better understanding of customer requirements leads to a better understanding of product function.†

## 3. Refine the Scope

The problem that has been identified often requires more time than the workshop schedule permits. In these cases, the problem should be rescoped to ensure that the most important elements are examined during the workshop. Plans for continuing the effort on the balance of the problem can be made at the end of the workshop.

Once the scope is determined and the final set of facts are collected from the best possible data sources, targets for improvement and evaluation factors should be reexamined and finalized. The study sponsor should approve any changes.

---

* Adapted from Kenneth Crow, *Customer-Focused Development with QFD* (Palos Verdes, CA: DRM Associates, 2002). Available: http://www.npd-solutions.com/qfd.html. Additional articles can be found in Robert A. Hunt, and Fernando B. Xavier, "The Leading Edge in Strategic QFD," *International Journal of Quality & Reliability Management* 20, no. 1 (2003): 56–73.

† Henry A Ball, "Value Methodology—The Link for Modern Management Improvement Tools," in *SAVE International 43rd Annual Conference Proceedings* (Scottsdale, AZ, June 8–11, 2003).

## C. Function Analysis Phase*

The Function Analysis phase identifies the most beneficial areas for study. The analytical efforts in this phase form the foundation of the job plan. The disciplined use of function analysis distinguishes the value methodology from other improvement methods. The following subsections describe the activities during the Function Analysis phase.†

### *1. Determine the Functions*

For the product or process under study, this activity encompasses determining forty to sixty functions that are performed by the product, the process, or any of the parts or labor operations. Functions are defined for every element of the product or process that consumes resources. The functions are typically recorded on adhesive-backed cards for later manipulation.

A function is defined as "the original intent or purpose that a product, service, or process is expected to perform."‡ Unstructured attempts to define the function(s) of a product or process will usually result in several concepts described in many words. Such an approach is not amenable to quantification. In VE, a function must be defined by two words: an active verb and a measurable noun:

- The verb should answer the question, "What does it do?" For example, it may generate, shoot, detect, emit, protect, or launch. This approach is a radical departure from traditional cost-reduction efforts because it focuses attention on the required action rather than the design. The traditional approaches ask the question, "What is it?" and then concentrate on making the same item less expensive by answering the question, "How do we reduce the cost of this design?"

---

* Some material in this section was adapted from information in US Army, *Value Engineering Program Management Guide*. US Army Materiel Command Pamphlet 11-3, 1986.

† These activities are adapted from SAVE International, *Function: Definition and Analysis* (October 1998), http://www.value-eng.org/pdf_docs/monographs/funcmono.pdf. They are consistent with those listed in SAVE International, *Value Standard and Body of Knowledge* (SAVE International Standard, June 2007), http://www.scribd.com/doc/15563084/Value-Standard-and-Body-of-Knowledge.

‡ SAVE International, *Value Standard and Body of Knowledge* (SAVE International Standard, June 2007), 28, http://www.scribd.com/doc/15563084/Value-Standard-and-Body-of-Knowledge.

- The noun answers the question, "What does it do this to?" The noun tells what is acted upon (e.g., electricity, bullets, movement, radiation, facilities, or missiles). It must be measurable or at least understood in measurable terms since a specific value must be assigned to it during the later evaluation process that relates cost to function.

A measurable noun, together with an active verb, provides a description of a work function (e.g., generate electricity, shoot bullets, detect movement, and so forth).

A work function establishes quantitative statements. Functional definitions containing a verb and a nonmeasurable noun are classified as *sell functions*. They establish qualitative statements (e.g., improve appearance, decrease effect, increase convenience, and so forth). Providing the correct level of function definition is important. For example, the function of a water service line to a building could be stated as "provide service." *Service*, not being readily measurable, is not amenable to determining alternatives. On the other hand, if the function of the line was stated as "conduct fluid," the noun in the definition is measurable, and the alternatives that are dependent upon the amount of fluid being transported can be readily determined.

Defining a function in two words, a verb and a noun, is known as *two-word abridgment*. The advantages of two-word abridgement are as follows:

- It forces brevity. If a function cannot be defined in two words, insufficient information is known about the problem, or the segment of the problem being defined is too large.
- It avoids combining functions and defining more than one simple function. By using only two words, the problem is broken down into its simplest element.
- It aids in achieving the broadest level of dissociation from specifics. When only two words are used, the possibility of faulty communication or misunderstanding is minimized.
- It focuses on function rather than on the item.
- It encourages creativity.
- It frees the mind from specific configurations.
- It enables the determination of unnecessary costs.
- It facilitates comparison.

## 2. Classify the Functions

The second major activity in the Function Analysis phase is to group the functions into two categories: basic and secondary.

The basic function is the intent and purpose of a product or process and answers the question, "What must it do?" Basic functions have or use value. A basic function defines the specific purpose(s) for which a product, facility, or service exists and conveys a sense of *need*.*

A product or process can possess more than one basic function, determined by considering the user's needs. A non-load-bearing exterior wall might be initially defined by the function description "enclose space." However, further function analysis determines that, for this particular wall, two basic functions are more definitive than the initial one: "secure area" and "shield interior." Both functions answer the question, "What does it do?"

Secondary functions answer the question "What else does it do?" Secondary functions are support functions and usually result from the particular design configuration. Generally, secondary functions contribute greatly to cost and may or may not be essential to the performance of the primary function. They support the basic function and result from the specific design approach used to achieve the basic function.†

As methods or design approaches to achieving the basic function are changed, secondary functions can also change. Three kinds of secondary functions are as follows:

1. **Required secondary functions.** These functions are necessary in a product or project to perform the basic function. For example, battery-operated flashlights and kerosene lanterns perform the basic function of producing light. A required secondary function, however, is to "conduct current" while the equivalent secondary function in the lantern is to "conduct fluid."
2. **Aesthetic secondary functions.** These functions add beauty or decoration to the product or project and are generally associated with *sell functions*. For example, the colors of paint available for a car could be an aesthetic secondary function.

---

* SAVE International, *Value Standard and Body of Knowledge*.

† SAVE International, *Value Standard and Body of Knowledge*.

3. **Unwanted secondary functions.** These functions, by definition, are not wanted while the product is performing the basic or secondary function(s). For example, while the kerosene lantern performs the basic function of producing light, an unwanted secondary function is that it "produces odor."*

Secondary functions that lend esteem value (convenience, user satisfaction, and appearance) are permissible only if they are necessary to permit the design or item to work or sell. These functions sometimes play an important part in the marketing or acceptance of a design or product. VE separates costs required for basic function performance from those incurred for secondary functions to eliminate as many non-value-added secondary functions as possible, improve the value of the remaining functions, and still provide the appeal necessary to permit the design or product to sell.

## *3. Develop Function Relationships*

Two principal techniques have been developed to create a better understanding of function relationships: a function hierarchy logic model and the Function Analysis System Technique (FAST).† This document concentrates on the classical FAST approach and the use of the FAST diagram.‡ FAST was developed by Charles W. Bytheway of the Sperry Rand Corporation and introduced in a paper presented at the 1965 National Conference of the

---

* James D. Bolton, Don J. Gerhart, and Michael P. Holt, *Value Methodology: A Pocket Guide to Reduce Cost and Improve Value through Function Analysis* (Lawrence, MA: GOAL/QPC, 2008), 46.

† These two approaches are described on an overview basis and illustrated using the same project in SAVE International, *Function Relationships: An Overview* (SAVE International Monograph, 1999), http://www.value-eng.org/pdf_docs/monographs/funcrelat.pdf.

‡ In addition to classical FAST, there are technical FAST and customer FAST. Technical FAST and customer FAST follow slightly different rules and formats. Additional information about the Function Hierarchy Logic model can be found in SAVE International, Function Logic Models (n.d.), http://www.value-eng.org/pdf_docs/monographs/funclogic.pdf. The equivalent publication on FAST is SAVE International, *Functional Analysis Systems Techniques—The Basics*, SAVE International Monograph (n.d.), http://www.value-eng.org/pdf_docs/monographs/FAbasics.pdf. The Army has published some FAST training material: *Function Analysis System Technique (FAST) Student Guide*, prepared by Nomura Enterprise, Inc., and J. J. Kaufman Associates, Inc., for the U.S. Army Industrial Engineering Activity, Rock Island, Illinois. The approach outlined in this section most closely follows J. Jerry Kaufman, *Value Engineering for the Practitioner* (Raleigh, NC: North Carolina State University, 1990).

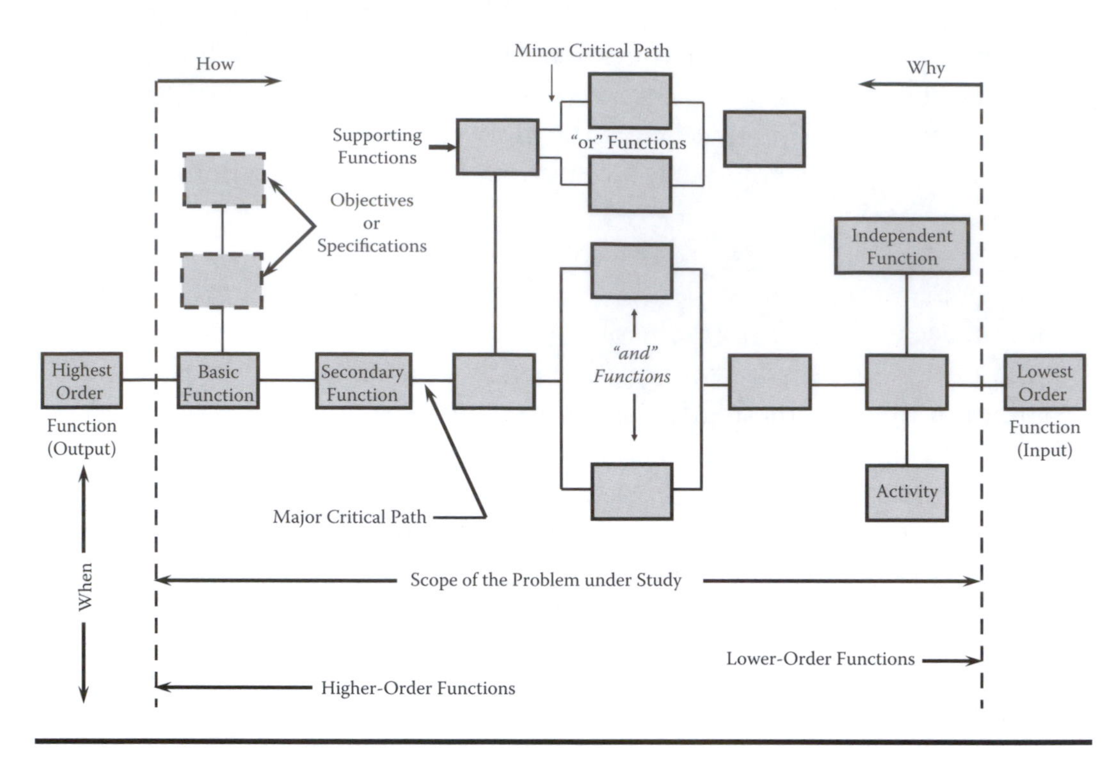


**Figure 2.2 Illustrative Classical FAST diagram.**

Society of American Value Engineers in Boston. Since then, FAST has been widely used by government agencies, private firms, and VE consultants. FAST is particularly applicable to a total project, program, or process requiring interrelated steps or a series of actions. Figure 2.2 illustrates a classical FAST diagram.

The basic classical FAST steps are as follows:

- **Step 1:** Determine the highest-order function. "The objective of the value study is called the Highest-Order Function(s) and is located to the left of the basic function(s) and outside the left scope line"* Determining the highest-order function is not always an easy process. For instance, the most offered highest-order function for a cigarette lighter is "lights cigarettes." This characterization, however, immediately raises the obvious question, "What about pipes and cigars?" An alternative might then be "generates flame." However, the electrical resistance lighter in a car only "emits energy." The thought process must focus in either one

* Robert B. Stewart, *Fundamentals of Value Methodology* (Bloomington, IN: Xlibris Corporation, 2005), 182.

direction or another to develop a multiplicity of two-word abridgements from which one or more levels can be chosen as the level of the basic functions to be studied.

- **Step 2:** Identify the basic functions. Select the basic functions that directly answer the question, "How does the product or process perform the highest-order function?" If all direct answers are not among the existing basic functions, create a new one. All of these basic functions should be included in the first column to the right of the higher-order function.
- **Step 3:** Expand the FAST diagram. Keep asking how the function is performed from the viewpoint of a user. Most answers will be found among the existing functions. Add second, third level, and lesser functions as needed to the right of the basic functions, but do not expand a function unless the "how" question is answered by two or more functions. Repeating the "how" question in this way is sometimes called the *ladder of abstraction* method. It is a thought-forcing process. Because using more than one definition can generate more creative ideas, this approach leads to greater fluency (more ideas), greater flexibility (variety of ideas), and improved function understanding of the problem. It generates critical paths for achieving the basic functions.
- **Step 4:** Identify the supporting functions. Supporting functions do not depend on another function. They are placed above a critical path and usually are needed to achieve the performance levels specified for the critical path function they support. The supporting functions above the critical path and the activities below the critical path are the result of answering the "when" question for a function on the critical path. A supporting function can have its own minor critical path.
- **Step 5:** Verify the FAST diagram. The FAST diagram is verified by driving one's thinking up the ladder of abstraction. Asking "why" raises the level, making the function description more general. In practice, the desired level is one that makes possible the largest number of feasible alternatives. Since the higher levels are more inclusive and afford more opportunities, the desired level is the highest level that includes applicable, achievable alternatives. A practical limit to the *why* direction is the highest level at which the practitioner is able to make changes. If the level selected is too low, alternatives can be restricted to those that resemble the existing design. If the level selected is too high, achievable alternatives can be obscured, and alternatives that are beyond the scope of effort might be suggested. FAST is generally used to understand a

problem, issue, or opportunity. However, developing a FAST diagram can be a difficult and time-consuming effort, but the decision to use a FAST diagram should be based on an understanding of the problem. The following broad considerations apply to such a decision:
- The more complicated the situation, the more useful a FAST diagram will be.
- If the situation is not well understood, a FAST diagram should be used.
- If there are more than three stakeholders that need to come to a common understanding of the situation, a FAST diagram should be used.
- If during the initial function analysis it is discovered that there are multiple secondary functions (particularly if they are codependent), a FAST diagram should be used.
- If the project is being carried out by a single individual, a FAST diagram should not be used unless the individual is already skilled at the technique.
- A FAST diagram should not be used if no one in the group has performed a Function Analysis before.
- If the scope is narrow and constrained, a FAST diagram may be necessary.

## *4. Estimate the Cost of Performing Each Function*

All VE studies include some type of economic analysis that identifies areas of VE opportunity and provides a monetary base from which the economic impact can be determined. The prerequisite for any economic analysis is reliable and appropriate cost data. Consequently, the VE study should use the services of one or more individuals who are skilled in estimating, developing, and analyzing cost data. The cost of the original or present method of performing the function (i.e., the cost for each block of the FAST diagram) is determined as carefully and precisely as possible given the time constraints for preparing the estimate.

The accuracy of a cost estimate for a product depends on

- the maturity of the item,
- the availability of detailed specifications and drawings, and
- the availability of historical cost data.

Similarly, the accuracy of a cost estimate for a service depends on

- the people involved,
- the time spent performing the service,
- the waiting time, and
- the direct, indirect, and overhead labor and material costs.

In some cases, a VE study will involve both products and process.

## 5. Determine the Best Opportunities for Improvement

The objective of this activity is to select functions for continued analyses. It is often accomplished by comparing function worth to function cost, where value is defined by the ratio of worth to cost (or cost to worth).* *Function worth* is defined as the lowest cost to perform the function without regard to consequences.

Thus, the use of function worth focuses the VE effort on those functions that will be most worthwhile and provides a reference point to compare alternatives. It can even be used as a psychological incentive to discourage prematurely stopping the VE effort before all of the alternatives are considered.

Determining the worth of every function is usually not necessary. Cost data aid in determining the priority of effort. Because significant savings potential in low-cost areas may not be a worthwhile pursuit and high-cost areas may be indicative of poor value, the latter are prime candidates for initial function worth determinations. Costs are usually distributed in accordance with Pareto's law of maldistribution: a few areas, "the significant few," (generally 20 percent or less) represent most (80 percent or more) of the cost. Conversely, 80 percent of the items, "the insignificant many," represent only 20 percent of total costs. Figure 2.3 illustrates this relationship.

A technique for developing the worth of functions, conceived in the early days of VE and still effective today, Pareto's law of maldistribution compares the selected function to the simplest method or product that can be imagined to achieve the same result. One increasingly popular technique for

* In practice, determining function worth is often difficult. As an alternative, total function cost can be distributed in a matrix whose rows are the functions and whose columns are components of a product or departments in a service or process scenario. Best opportunities for improvement are sought among the highest-cost functions. The relative worth of components can also be inferred from a customer's relative value of design functions. An interesting example of using QFD to do this can be found in K. Ishii and S. Kmenta, *Life-cycle Cost Drivers and Functional Worth*, Project Report for ME317: Design for Manufacturing, Department of Mechanical Engineering (Palo Alto, CA: Stanford University, n.d.).

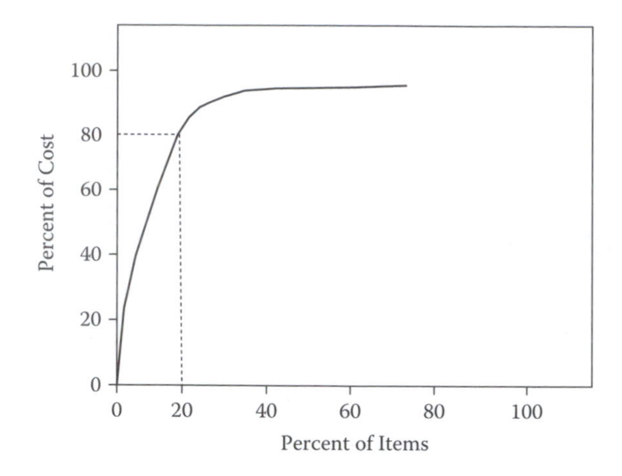


**Figure 2.3 Pareto's law of maldistribution.**

assigning worth to functions ascertains the primary material cost associated with the function.*

The value calculation can be done in many ways. For example, some workshop facilitators use a ratio of *percent relative importance* to *percent of cost*. In this approach, all functions are evaluated pairwise, with different numbers assigned to reflect the relative importance of the two functions being compared (e.g., 3 may mean a large difference in importance, 1 may mean a small difference in importance). A relative importance is calculated for each function individually as the sum of the relative importance scores that function received when it was ranked higher than another function in the pairwise comparisons. The *percent relative importance* is calculated by converting the individual function's relative importance scores to a percentage of the total. The *percent of cost* is the cost of a function relative to the total cost of all functions.† There are other approaches. For example, Thomas Snodgrass‡ suggests an alternative approach based on high, medium, and low scores for function acceptance, function cost, and function importance.

Whatever approach is used, the best opportunities for improvement are determined by improving functions that have excessively low ratios of worth to cost (or high ratios of cost to worth). This ratio is referred to as the *value index*.

---

* SAVE International, *Function: Definition and Analysis* (October 1998), http://www.value-eng.org/pdf_docs/monographs/funcmono.pdf.

† A more complete description can be found in Arthur E. Mudge, *Value Engineering: A Systematic Approach* (Pittsburgh, PA: J. Pohl Associates, 1989), 68–74.

‡ Thomas J. Snodgrass, "Function Analysis and Quality Management," in *SAVE International 33rd Annual Conference Proceedings* (1993).

### 6. Refine Study Scope

The final activity in the Function Analysis phase refines the study scope to reflect any changes that have taken place.

## D. Creative Phase

The Creative phase develops ideas for alternative ways to perform each function selected for further study. The two approaches to solving a problem are *analytical* and *creative*. In the analytical approach, the problem is stated, and a direct, step-by-step approach to the solution is taken. An analytical problem frequently has only one solution that will work. The analytical approach should *not* be used in the Creative phase. The creative approach is an idea-producing process specifically intended to generate a number of solutions that solve the problem at hand. All solutions could work, but one is better than the others. It is the optimum solution among those available. Once a list of potential solutions is generated, determining the best value solution is an analytical process conducted in the latter phases of the job plan.

Creative problem-solving techniques are an indispensable ingredient of effective VE. By using the expertise and experience of the study team members, new ideas will be developed. The synergistic effect of combining the expertise and experience of all team members will lead to a far greater number of possibilities. The following subsections describe the activities during the Creative phase (also called the Speculation phase).

### 1. Discourage Creativity Inhibitors

For these activities to work well, the team must avoid mental attitudes that hinder creativity. The facilitator should point out creativity inhibitors to the team. Awareness of these inhibitors encourages people to overcome them. Parker identifies the following as common habitual, perceptual, cultural, and emotional blocks to creativity:*

- Habitual blocks
  - Continuing to use "tried and true" procedures even though new and better ones are available

---

* Donald E. Parker, *Value Engineering Theory*, rev. ed. (Washington, DC: The Lawrence D. Miles Value Foundation, 1998), 93.

  - Rejecting alternative solutions that are incompatible with habitual solutions
  - Lacking a positive outlook, lacking effort, conformity to custom, and reliance on authority
- Perceptual blocks
  - Failure to use all the senses for observation
  - Failure to investigate the obvious
  - Inability to define terms
  - Difficulty in visualizing remote relationships
  - Failure to distinguish between cause and effect
  - Inability to define the problem clearly in terms that will lead to the solution of the real problem
- Cultural blocks*
  - Desire to conform to proper patterns, customs, or methods
  - Overemphasis on competition or cooperation
  - The drive to be practical above all else, thus making decisions too quickly
  - Belief that all indulgence in fantasy is a waste of time
  - Faith only in reason and logic
- Emotional blocks
  - Fear of making a mistake or of appearing foolish
  - Fear of supervisors and distrust of colleagues
  - Too much emphasis on succeeding quickly
  - Difficulty in rejecting a workable solution and searching for a better one
  - Difficulty in changing set ideas (no flexibility) and depending entirely upon judicial (biased) opinion
- Inability to relax and let incubation take place

The following list adapted from Michel Thiry's "good idea killers" could also be used to make the team aware of attitudes to avoid:†

- It is not realistic.
- It is technically impossible.
- It does not apply.
- It will never work.

* Political blocks can also be included here.

† Michel Thiry, *Value Management Practice* (Newtown Square, PA: Project Management Institute, 1997), 57.

- It does not correspond to standards.
- It is not part of our mandate.
- It would be too difficult to manage.
- It would change things too much.
- It will cost too much.
- Management will never agree.
- We do not have time.
- We have always done it that way.
- We already tried it.
- We have never thought of it that way.
- We are already too far into the process.

The Creative phase does not necessarily identify final solutions or ideas ready for immediate implementation. It often simply provides leads that point to final solutions.

Beginning the Creative phase with a creativity-stimulating exercise can also be useful. J. Jerry Kaufman and James D. McCuish* report a threefold increase in ideas with the use of such a stimulus. They suggest using the Impossible Invention creativity exercise developed in the Massachusetts Institute of Technology (MIT) creativity lab in the 1960s. This 30-minute exercise consists of dividing the participants into three- or four-person teams. Each team then progresses through preliminary steps to select the three worst ways to perform the function without knowing why or the parameters that define *worst.* The objective of the exercise is for team members—as a team and as individuals—to experience how far beyond the teams' paradigm they can venture in an environment in which their self-esteem is protected.

## 2. Establish Ground Rules

The ground rules for creative idea generation, as adapted from Parker,† are summarized as follows:

* J. Jerry Kaufman and James D. McCuish, "Getting Better Solutions with Brainstorming," in *SAVE International 42nd Annual Conference Proceedings* (Denver, CO, May 5–8, 2002).

† Donald E. Parker, *Value Engineering Theory*, rev. ed. (Washington, DC: The Lawrence D. Miles Value Foundation, 1998), 96.

- Do not attempt to generate new ideas and judge them at the same time. Reserve all judgment and evaluation until the Evaluation phase.
- Focus on quantity, not quality. Generate a large quantity of possible solutions. As a goal, multiply the number of ideas produced in the first rush of thinking by five or even ten.
- Seek a wide variety of solutions that represent a broad spectrum of attacks on the problem. The greater the number of ideas conceived, the greater likelihood of an alternative that leads to better value.
- Freewheeling is welcome. Deliberately seek unusual ideas.
- Watch for opportunities to combine or expand ideas as they are generated. Include them as new ideas. Do not replace anything.
- Do not discard any ideas, even if they appear to be impractical.
- Do not criticize or ridicule any ideas. (Criticism could be discouraged, for example, by maintaining a criticizer list or imposing a mock penalty on criticizers.)

## *3. Generate Alternative Ideas*

In this phase of the study, generating a free flow of thoughts and ideas for alternative ways to perform the functions—not how to design a product or service—is important. While creativity tools are available for problem-solving situations, no specific combination of techniques is prescribed for all VE projects, and the degree to which they should be used is not predetermined. The selection of specific techniques and the depth to which they are used are primarily matters of judgment and vary according to the complexity of the subject under review.

The following list of idea-generation techniques describes some commonly used approaches in the VE context:*

- **Brainstorming.** Brainstorming is a free-association technique that groups use to solve specific problems by recording spontaneous ideas generated by the group. It is primarily based on the premise that one idea suggests others, which suggest even more. An individual can brainstorm, but experience has shown that a group can generate more ideas collectively than the same number of persons thinking individually. Roger B. Sperling has suggested combining group and individual

* Some of the following material was adapted from information in US Army. *Value Engineering Program Management Guide*. US Army Materiel Command Pamphlet 11-3, 1986.

brainstorming.* He found that after the group brainstorming was complete, individual brainstorming can generate additional ideas of comparable quality.

- **Gordon technique.** The Gordon technique is closely related to brainstorming. The principal difference is that no one except the group leader knows the exact nature of the problem under consideration. This difference helps avoid the premature ending of the session or egocentric involvement. A participant may cease to produce additional ideas or devote energy only to defending an idea if he/she is convinced that one of the ideas already proposed is the best solution to the problem. Selecting a topic for such a session is more difficult than selecting a topic for a brainstorming session. The subject must be closely related to the problem at hand, but its exact nature must not be revealed until the discussion is concluded.
- **Checklist.** The checklist technique generates ideas by comparing a logical list of categories with the problem or subject under consideration. Checklists range from the specialized to the extremely general.
- **Morphological analysis.** Morphological analysis is a structured, comprehensive system for methodically relating problem elements to develop new solutions. In this approach, the problem is defined in terms of its dimensions or parameters, and a model is developed to visualize every possible solution. Problems with too many parameters rapidly become intractable.
- **Attribute listing.** The attribute listing approach lists all of the various characteristics of a subject first and then measures the impact of changes. By so doing, new combinations of characteristics (attributes) that will better fulfill some existing need can be determined.
- **Input–output technique.** The input–output technique establishes output, establishes input as the starting point, and varies combinations of input/output until an optimum mix is achieved.
- **Theory of Inventive Problem Solving (TRIZ).** TRIZ (stands for the Russian *Teoriya Resheniya Izobretatelskikh Zadatch*) is a management tool that will be used more frequently with greater awareness of its capabilities. The methods and tools are embodied in five steps: problem documentation and preliminary analysis, problem formulation, prioritization of directions for innovation, development of concepts, and

* Roger B. Sperling, "Enhancing Creativity with Pencil and Paper," in *SAVE International 39th Annual Conference Proceedings* (San Antonio, TX, June 27–30, 1999), 284–289.

> evaluation of results. C. Bernard Dull points out that both VE and TRIZ have strengths and weaknesses.[*] Combining these two problem-solving methodologies can create synergies that lead to more robust and comprehensive results, especially for more technically complex projects where the added benefit is worth the effort. He suggests that integrating TRIZ into the VE job plan is easier than integrating VE into the TRIZ job plan. Dana W. Clarke goes into greater detail in the Creative phase by suggesting how TRIZ can be used to augment traditional brainstorming.[†] Ball supports Clarke's conclusion, stating: "This is a much more intensive method of identifying potential solutions than generally used in a VM [Value Management] study."[‡]

When using any one of these techniques, the team reviews the elements of the problem several times. If possible, new viewpoints should be obtained by discussing the problem with others. Different approaches should be used if one technique proves to be ineffective.[§] However, before rejecting any possible solutions, one effective strategy allows the team to take a break to allow time for subconscious thought on the problem while consciously performing other tasks.

## E. Evaluation Phase

The Evaluation phase selects and refines the best ideas to develop into specific value improvement recommendations. Ultimately, the team should present the decision maker with a small number (e.g., fewer than six) of choices. In the Creative phase, a conscious effort was made to prohibit judgmental thinking because it inhibits the creative process. In the Evaluation phase, all the alternatives must be critically assessed to identify the best opportunities for value improvement. This phase is not the last

---

[*] C. Bernard Dull, "Comparing and Combining Value Engineering and TRIZ Techniques," in *SAVE International 39th Annual Conference Proceedings* (San Antonio, TX, June 27–30, 1999), 71–76.

[†] Dana W. Clarke, Sr., "Integrating TRIZ with Value Engineering: Discovering Alternative to Traditional Brainstorming and the Selection and Use of Ideas," in *SAVE International 39th Annual Conference Proceedings* (San Antonio, TX, June 27–30, 1999), 42–51.

[‡] Henry A Ball, "Value Methodology—The Link for Modern Management Improvement Tools," in *SAVE International 43rd Annual Conference Proceedings* (Scottsdale, AZ, June 8–11, 2003).

[§] Some work has been done on a systematic approach for moving between creative methodologies. See Donald Hannan, "A Hybrid Approach to Creativity," in *SAVE International 41st Annual Conference Proceedings* (Fort Lauderdale, FL, May 6–9, 2001).

chance to defer ideas. A detailed cost–benefit analysis conducted in the Development phase leads to the final set of choices presented to the decision maker. The following subsections describe the activities during the Evaluation phase.

### *1. Eliminate Low-Potential Ideas*

Ideas that are not feasible, too hard, not promising, or do not perform the basic function should be eliminated. A useful approach to this activity is to classify the ideas into three categories:

- **Yes.** These ideas appear to be feasible and have a relatively high probability of success.
- **Maybe.** These ideas have potential but appear to need additional refinement or work before they can become proposals.
- **Not Now.** These ideas have little or no potential at this time.

At this point, eliminate only the *not now* ideas.

### *2. Group Similar Ideas*

The remaining ideas are grouped into several (three or more) subject-related categories and examined to determine if they should be modified or combined with others. Sometimes, the strong parts of two different ideas can be developed into a winning idea. In other cases, several ideas can be so similar that they can be combined into a single all-encompassing idea. Some workshops employ a *forced relationships* technique that deliberately attempts to combine ideas from the different subject-related categories to discover new, innovative alternatives.

### *3. Establish Idea Champions*

The remaining activities in this phase are designed to prioritize the ideas for further development. An *idea champion* is a study team member who will serve as an idea's proponent throughout the prioritization process. If an idea has no champion, it should be eliminated at this point.

## 4. List the Advantages and Disadvantages of Each Idea

The advantages and disadvantages of each idea are identified along with the ease of change, cost, savings potential, time to implement, degree to which all requirements are met, and likelihood of success. All of the effects, repercussions, and consequences that might occur in trying to accomplish a solution should be anticipated.

Useful suggestions include how to overcome the disadvantages. No matter how many advantages an idea has, disadvantages that cannot be overcome may lead to its rejection.

## 5. Rank the Ideas

A set of evaluation criteria should be developed to judge the ideas, using the factors considered when listing advantages and disadvantages (e.g., cost, technical feasibility, likelihood of approval, time to implement, and potential benefit). The ideas should be ranked according to the criteria that have been developed. No idea should be discarded, and all ideas should be evaluated as objectively as possible. Ratings and their weights are based on the judgment of the people performing the evaluation. Techniques such as evaluation by comparison, numerical evaluation, or team consensus can be used. Simplified decision analysis techniques such as QFD can also be applied. Yuh-Huei Chang and Ching-Song Liou suggest using a simplified risk identification and analysis process to evaluate the performance of alternatives and combining these results with criteria weights to determine the best alternatives for further development.[*]

This initial analysis will produce a shorter list of alternatives, each of which has met the evaluation standards set by the team. At this point in the Evaluation phase, adapting an idea suggested by John D. Pucetas for the Creative phase might be useful. Pucetas recommends using *force field* analysis, which evaluates helping and hindering forces in the pursuit of a product, to "measure the sensitivity of the VE team regarding controversial project issues."[†] For the higher-ranked ideas, the VE team should suggest ways to reduce the disadvantages and enhance the advantages. This exercise can lead to the following potential benefits:

[*] Yuh-Huei Chang and Ching-Song Liou, "Implementing the Risk Analysis in Evaluation phase to Increase the Project Value," in *SAVE International 45th Annual Conference Proceedings* (San Diego, CA, June 26–29, 2005).

[†] John D. Pucetas, "Keys to Successful VE Implementation," in *SAVE International 38th Annual Conference Proceedings* (Washington, DC, June 14–17, 1998), 340.

- Ideas can be revised to improve their potential for success.
- Insight into implementation issues can be obtained from the suggested ways to reduce the disadvantages.
- Insight into the acceptability of the idea and the likelihood of management approval can be derived from suggested ways to enhance the advantages.

This approach can serve as a basis for distinguishing among the higher-ranked ideas (i.e., reranking the ideas) and as a consequence, simplifying and strengthening the procedure for selecting ideas for further development.

### *6. Select Ideas for Further Development*

Typically, a cutoff point is established for identifying ideas for further development. If a natural break occurs in quantitative evaluation scores, a cutoff point may be obvious. If only qualitative evaluation scores are used or if quantitative scores are close, a more refined ranking scheme may be needed to make the selection. However, if several alternatives are not decisively different at this point, they should be developed further.

Alternatives with the greatest value potential will normally be among those selected. If that is not the case, those ideas should be reexamined to determine whether they should be developed further. Retaining at least one idea from each of the subject-related categories used to group ideas at the beginning of the Evaluation phase is also useful.

## F. Development Phase

The Development phase determines the "best" alternative(s) for presentation to the decision maker. In this phase, detailed technical analyses are made for the remaining alternatives. These analyses form the basis for eliminating weaker alternatives. The following subsections describe the activities during the Development phase.

### *1. Conduct a Life-Cycle Cost Analysis*

A life-cycle cost analysis ranks all remaining alternatives according to an estimate of their life-cycle cost-reduction potential relative to the status quo. Cost estimates must be as complete, accurate, and consistent as possible to minimize the possibility of error in assessing the relative economic potential of the alternatives. Specifically, the method used to cost the status quo should also be used to cost the alternatives.

All costs should be identified. For the originating organization, costs may include the following:

- New tools or fixtures
- Additional materials
- New assembly instructions
- Changes to plant layout and assembly methods
- Revisions to test and/or inspection procedures
- Retraining assembly, test, or inspection personnel
- Reworking parts or assemblies to make them compatible with the new design
- Tests for feasibility

Other costs that are not normally incurred by the originating activity, but should be considered, include the following:

- Technical and economic evaluation of proposals by cognizant personnel
- Prototypes
- Testing the proposed change
- Additional equipment that must be provided
- If applicable, retrofit kits (used to change design of equipment already in field use)
- Installation and testing of retrofit kits
- Changes to engineering drawings and manuals
- Training personnel to operate and maintain the new item
- Obtaining new and deleting obsolete stock numbers
- Paperwork associated with adding or subtracting items from the supply system
- Maintaining new parts inventory in the supply system (warehousing)
- Purging the supply system of parts made obsolete by the change
- Changing contract work statements and specifications to permit implementation of the proposal

Determining the precise cost associated with a proposed change is not always possible. For example, the actual cost of revising, printing, and issuing a page of a maintenance manual is difficult to obtain. Nevertheless, this charge is a recognized item of cost because the manual must be changed if the configuration of the item is changed. One common practice uses a schedule of surcharges to cover areas of cost that defy precise determination. Such a schedule is usually based on the average of data obtained from various sources.

Comparing alternatives using a *constant dollar* analysis instead of a *current dollar* approach is easier. It permits labor and material cost estimates to be based on current operational and maintenance data and eliminates the need to figure out how they would inflate in some future year. The net present worth of each of the alternatives should be calculated but only after management agrees on two factors:

- **The discount rate to be used.** This figure is the difference between the inflation rate assumed and the time value of money (interest rate).
- **The length of the life cycle.** This measurement is the number of years of intended use or operation of the object being studied.

The Office of Management and Budget (OMB) provides annual guidance on appropriate discount rates.* Normally, the Department of Defense (DOD) allows a period of fifteen to twenty years as a reasonable life cycle. However, a program or a command may have different guidance for a particular situation.

## 2. Determine the Most Beneficial Alternatives

In evaluating the alternative, the VE team should consult personnel who have knowledge about the item's function, operational constraints, dependability, and requirements. Technical problems related to design, implementation, procurement, or operation must also be determined and resolved. Certain key questions should be answered as part of this effort:

- What are the life-cycle savings?
- Do the benefits outweigh the costs?
- What are the major risks?
- How can the risks be mitigated?
- Are any technical issues outstanding?

---

* OMB Circular A-94, "Guidelines and Discount Rates for Benefit-Cost Analysis of Federal Programs" (Washington, DC: 1992).

If more than one alternative offers a significant savings potential, the common practice is to recommend all of them. One becomes the primary recommendation, and the others are alternative recommendations, usually presented in decreasing order of saving potential. Other nonquantified benefits should also be considered.

The VE team should consult personnel who have knowledge about the item's function, operational constraints, dependability, and requirements. Technical problems related to design, implementation, procurement, or operation must also be determined and resolved.

### *3. Develop Implementation (Action) Plans*

The implementation plan for each alternative should include a schedule of the required implementation steps; identify who will execute the plan; specify the resources required, the approval process, the necessary documents, the timing requirements, the coordination required; and so forth. The team must anticipate problems relating to implementation and propose specific solutions to each. Discussions with specialists in relevant areas are particularly helpful in solving such problems.

When needed, testing and evaluation should be planned for and scheduled during the recommended implementation process. Occasionally, concurrent testing of two or more proposals allows a significant reduction in the implementation investment. Also, significant reductions in testing cost can often be achieved by scheduling tests into other test programs scheduled within the desirable timeframe—especially when items to be tested are a part of a larger system also being tested. However, care must be exercised during combined testing to prevent masking the feasibility of one concept by the failure of another.

## G. Presentation Phase

The purpose of the Presentation phase is to obtain a commitment to follow a course of action and initiate an alternative. The VE team makes a presentation to the decision maker (or study sponsor) at the conclusion of the workshop. This presentation is normally the first step (not the last step) in the approval process. Typically, a decision to implement is not made at the time of the briefing.

Additional steps include the following:

- Answering follow-on questions
- Collecting additional data
- Reviewing supporting documentation
- Involving other decision makers

The sole activity during this phase involves preparing a presentation to encourage commitment. An oral presentation can be the cornerstone to selling a proposal. It should have an impact and continue the process of winning management and other stakeholder support. This presentation gives the VE team a chance to ensure that its written proposal is correctly understood and that proper communication exists between the parties concerned. The presentation's effectiveness will be enhanced if

- the entire team is present and introduced;
- the presentation lasts no longer than twenty minutes, with time for questions at the end;
- the presentation is illustrated using mockups, models, slides, vu-graphs, or flip charts; and
- the team has prepared sufficient backup material to answer all questions posed during the presentation.

The presentation itself should

- describe the workshop objectives and scope;
- identify the team members and recognize their contributions;
- describe the *before* and *after* conditions for each alternative;
- present the costs and benefits, advantages and disadvantages, and impact of each alternative;
- identify strategies to overcome roadblocks;
- demonstrate the validity of the data sources; and
- suggest an action plan and implementation schedule.

The most successful strategies to improve the probability of success and reduce the time required for acceptance and implementation of proposals appear to be the following:

- **Consider the reviewer's needs.** Terminology appropriate to the training and experience of the reviewer should be used. Each proposal is usually directed toward two audiences: (1) the technical authority that requires sufficient technical detail to demonstrate the engineering feasibility of the proposed change and (2) the administrative reviewers for whom the technical details can be summarized, but for whom the financial implications (cost and likely benefits) are emphasized. Long-range effects on policies, procurement, and applications are usually more significant to the administrator than to the technical reviewer.
- **Address risk.** Decision makers are often more interested in the risk involved in making a decision than the benefits or value that might be achieved by the decision. Decision-making risk should not be confused with technical risk. Decision-making risk encompasses the uncertainty and complexity generated from making a change. Therefore, the organizational culture and behavior should be considered when characterizing the recommendation.
- **Relate benefits to organizational objectives.** A proposal that represents advancement toward some approved objective is most likely to receive favorable consideration from management. Therefore, the presentation should exploit all of the advantages that a proposal can offer toward fulfilling organizational objectives and goals. When reviewing a proposal, the manager normally seeks either lower total cost of ownership or increased capability at the same or lower cost. The objective may be not only savings, but also the attainment of some other mission-related goal of the manager.
- **Show collateral benefits of the investment.** Often, VE proposals offer greater benefits than the cost improvement specifically identified. Some of the benefits are collateral in nature and can be difficult to quantify. Nevertheless, collateral benefits should be included in the proposal. The likelihood of the proposal's acceptance is improved when all of its collateral benefits are clearly identified and completely described.

The Presentation phase should end with a list of actions leading to approval:

- Preparation and submission of a final workshop report with all the necessary supporting documentation
- Briefings to other key stakeholders
- A schedule for a follow-up meeting to approve the proposal

## H. Implementation Phase

The purpose of the Implementation phase is to obtain final approval of the proposal and facilitate its implementation. Throughout this phase, the team should be mindful of factors that contribute to successful change. According to R. A. Fraser:

> The VE/VA techniques provide an excellent method for planned and managed change. However, even when the job plan is applied well, challenges to the change process occur due to individual differences and human interpretation. At each stage of the change process, a number of varying responses may be expected from individuals involved throughout the organization. These responses range from active support to resistance. One of the approaches that has demonstrably improved the chances for success of the planned change and reduced reactive resistance is to let people in on the action—to participate in the decision-making process.[*]

Fraser also notes the five factors David A. Kolb and Richard E. Boyatzis identified as being most related to achieving a goal: awareness, expectation of success, psychological safety, measurability of the change goal, and self-controlled evaluation.[†]

VE is ideally suited to meeting these challenges. The following subsections describe the activities during the Implementation phase.

### 1. Prepare a Written Report

The oral presentation of study results is most helpful to the person who is responsible for making the decision; however, it should never replace the written report. A written report normally demands and receives a written reply, whereas an oral report can be forgotten and overlooked after it is presented. In the rush to conclude a project, promote a solution, or avoid the effort of writing a report, many proposals fail to materialize because the

---

[*] R. A. Fraser, "The Value Manager as Change Agent or How to be a Good Deviant," in *SAVE International Annual 24th Conference Proceedings* (Sacramento, CA, May 6–9, 1984), 199–203. The acronym *VA* in the quote means *Value Analysis*, which is synonymous with VE.

[†] David A. Kolb and Richard E. Boyatzis, "Goal Setting and Self-Directed Behavior Change," in *Organizational Psychology: A Book of Readings*, ed. David A. Kolb, Irwin M. Rubin, James M. McIntyre (Englewood Cliffs, NJ: Prentice-Hall, 1979).

oral presentation alone is inadequate. The systematic approach of the VE job plan must be followed to conclusion and should include the meticulous preparation of a written report.

Like any other well-prepared report, this final report should

- satisfy questions the decision maker is likely to ask,
- provide assurance that approval would benefit the organization,
- include sufficient documentation to warrant a favorable decision with reasonable risk factors (both technical and economic), and
- show how performance is not adversely affected.

Well-prepared teams get a head start on the final report by documenting the progress between phases. For example, before the Development phase, the team should develop documentation detailing what is proposed, to what extent the idea meets the criteria established in the Orientation phase, risks, investment costs, expected savings, and so forth for each surviving idea.

The final report should be accompanied by a team letter that summarizes the recommendation and action plan and requests action from the sponsor. It should be sent with the report to all stakeholders.

## 2. Enhance the Probability of Approval

Approval of a proposal involves change to the status quo. Because of this or other pressing priorities, a manager may be slow in making a decision.

The manager who makes an investment in a VE study expects to receive periodic progress reports before a final decision is made. Regular reporting helps ensure top management's awareness of, support for, and participation in any improvement program. Therefore, the change should be discussed with the decision makers or their advisors before and after the final report has been submitted. This practice familiarizes key personnel with impending proposals and enables a more rapid evaluation. Early disclosure can also serve to warn the originators of any objections to the proposal. This "early warning" will give the originators an opportunity to incorporate explanations and details into the final report to overcome the objections. These preliminary discussions often produce additional suggestions that improve the proposal and enable the decision maker to contribute directly.

Implementation depends on an expeditious approval by the decision makers in each organizational component affected by the proposal. The VE team members should serve as liaisons between decision makers and other stakeholders by preparing information that weighs the risks against the potential rewards and by identifying potential roadblocks and solutions.

Some organizations convene an implementation meeting with all stakeholders.* Once tentative decisions are made, this meeting is used to help everyone understand which proposals or modified proposals have been accepted or rejected or will be studied further. In some cases, the tentative decisions are changed based on clarification of a misunderstood assumption.

## 3. Monitor Progress

Implementation progress must be monitored just as systematically as the VE study. The VE team should ensure that implementation is actually achieved. A person could be given the responsibility of monitoring the deadline dates in the implementation plan and the process of obtaining any implementation funding.

## 4. Expedite Implementation

To minimize delays in the implementation process, the VE team should provide assistance, clear up misconceptions, and resolve problems that may develop in the implementation process. When possible, the VE team should prepare first drafts of the documents necessary to revise handbooks, the specifications, the change orders, the drawings, and the contract requirements. Such drafts help to ensure proper translation of the idea into action and serve as a baseline from which to monitor progress of final implementation. The VE team should review all implementation actions to ensure communication channels are open and that approved ideas are not compromised by losing their cost effectiveness or the basis for their original selection.

## 5. Follow-Up

The final activity of the Implementation phase includes several diverse tasks that foster and promote the success of subsequent VE efforts:

---

* Jill Ann Woller, "Value Analysis: An Effective Tool for Organizational Change," in *SAVE International 45th Annual Conference Proceedings* (San Diego, CA, June 26–29, 2005).

- Obtain copies of all completed implementation actions.
- Compare actual results with original expectations.
- Submit cost savings or other benefit reports to management.
- Submit technical cross-feed reports to management.
- Conduct a *lessons-learned* analysis of the study to identify problems encountered and recommend corrective action for the next study.
- Publicize accomplishments.
- Initiate recommendations for potential future VE studies on ideas evolving from the study just completed.
- Screen all contributors to the effort for possible receipt of an award and initiate recommendations for appropriate recognition.

# Chapter 3

# Lean Six Sigma (LSS) Methodology*

This chapter describes two methodological approaches to LSS. Some LSS proponents have asserted that no continuous process improvement methodology has "a more balanced approach or success than Lean Six Sigma."† In fact, the word *Lean*, when used as an adjective, often connotes a new and streamlined way of carrying out some activity using Lean principles. For example, Lean design has been defined as "the power to do less of what doesn't matter and more of what does matter."‡

The first approach to LSS is the Define, Measure, Analyze, Improve, and Control (DMAIC) methodology. It is by far the most common. The steps in the DMAIC process are described in Section A and diagrammed in Figure 3.1. The LSS variant Design for Six Sigma (DFSS) methodology is discussed in Section B.

## A. The DMAIC Methodology

### 1. Define Phase

The LSS methodology begins with the Define phase when a problem area is first recognized and opportunities to reduce waste or variation are explored.

* The material in this chapter was adapted from the Department of Defense (DOD) LSS Black Belt Course and the DOD LSS Champion Course as contained in the training page of https://www.us.army.mil/suite/page/596053. (This page is accessible to government employees and people they sponsor.)

† J. D. Sicilia, "Champion Training" (briefing), Office of the Secretary of Defense.

‡ Bart Huthwaite, *The Lean Design Solution* (Mackinac Island, MI: Institute for Lean Design, 2004).

| Define | Measure | Analyze | Improve | Control |
|---|---|---|---|---|
| What is the problem area? | What do we know about the problem? | What are the root causes? | How can we improve process performance? | How can we manage and sustain change? |
| Identify the Problem Area | Develop and Analyze the Process Maps | Identify Potential Root Causes | Brainstorm Improvement Ideas | Execute the Implementation Plan |
| Assess the Problem and Requirements for Process/Product Improvement | Prioritize Measurement Tasks | Prioritize Potential Root Causes | Generate Solutions to Make the Process Leaner | Establish Controls |
| Establish Goals | Identify Metrics | Analyze the High-Priority Root Causes | Determine Ways to Eliminate Waste | Take Corrective Actions (as Appropriate) |
| Form and Orient a Team | Develop a Data Collection Plan | | Determine Ways to Sustain Waste Eliminiation | |
| Develop Initial Process Maps | Collect Data | | Determine the Most Important Actions to Be Taken | |
| Create Plans for Overcoming Barriers, Communication, and Schedule | | | Create an Implementation Plan | |
| Finalize the Problem Statement with a Charter | | | | |

**Figure 3.1 The DMAIC process.**

The objective of the Define phase is to comprehensively examine a problem area by narrowing down and scoping the areas of deficiency. The Define phase entails identifying an area for improvement, developing a more detailed understanding of the associated process, identifying project goals, forming a team, developing initial process maps, identifying roadblocks and solutions, and finalizing a problem statement to guide project work.

The sponsor plays a vital role in the Define phase by communicating requirements, goals, and guidance to the team and steering the project and managing the budget. Other major players include the process owner who may also be the sponsor project champion, who serves as a go-between for the team and senior leadership and approves major decisions; an LSS Black Belt and/or LSS Master Black Belt who oversees and manages the project, provides expert guidance, and trains and prepares the team; and the team members. An effective Define phase will recognize a problem and work with a team to lay the groundwork for further analysis. The following subsections describe the activities during the Define phase.

## a. *Identify the Problem Area*

Identifying a problem area is the first step in LSS and process improvement to eliminate waste. This step can entail a broad and brief description of a problem based on observation; however, the problem must be quantifiable. A problem can be identified by anyone at any point in the process, either from a top-down or bottom-up perspective. Problem areas can include cost, productivity, time, defects, safety, or customer satisfaction potential. Corresponding indicators include cost overruns, inefficient use of resources, process overlaps, or an inadequate product.

Sponsor input is crucial in identifying a problem area. This input provides important insights about his/her and the process owner's needs, objectives, or specifics about a process that may not be obvious to an outside observer. A problem does not necessarily need to be identified first by the process owner, but initial findings and recommendations need to be communicated to the champion and customer as part of the identification process.

Part of the identification step is also developing an awareness of the benefits of improving the process by rectifying a problem. Benefit analysis helps by examining parts of the process that relay the greatest value and focusing on those areas that will increase productivity and capture the benefits.

### *b. Assess the Problem and Requirements for Product/Process Improvement*

Once a problem is identified, the next step is for the champion to work with the process owner and/or sponsor to develop a deeper understanding of the problem and how it fits into a larger process. Assessing the problem area entails moving to an additional level of detail to frame the way ahead and to develop a better understanding of the customer's needs. The assessment should identify what is critical to quality, cost, delivery, safety, customer satisfaction, or any sector within which the problem area falls. Questions for consideration include the following:

- How does the problem impact the entire process?
- Does the entire process have to be changed to address this specific problem?
- What is the most crucial part of the process?
- Can the deficiency be quantified?
- Can the targets be quantified?

This assessment identifies the specific issue for the LSS project and lays the groundwork for moving forward with a plan to improve the process.

The sponsor plays a critical role in this step. He/she provides requirements for process improvement, which must be measurable and relate directly to the product or service. All information about the problem area must be validated with the process owner. The sponsor and the champion will identify important relationships within the organization, the process and project's level of importance, and any additional requirements or constraints for the LSS team.

### *c. Establish Goals*

The goal of the previous steps was to identify and focus on a specific problem from which specific goals can be derived with the help of the sponsor. Tools for communicating with the customer include the following:

- **Likert scale.** The Likert scale measures the strength of agreement with a given statement about the process through a questionnaire to gauge attitudes.

- **Surveys.** Surveys provide specific questions to gauge the customer's concerns and requirements. However, participation is often lacking in surveys.
- **Interviews.** Interviews solicit candid feedback and provide the process owner with a direct voice; allow for a free flow exchange of ideas about what is wrong and what can be fixed; and establish face-to-face communication with the sponsor, which may be useful throughout the project.
- **Focus groups.** These groups establish a panel to answer questions about the product or process and solicit feedback directly from those involved in the process.

Project goals should follow the SMART guidelines: specific, measurable, achievable, relevant, and time-bound. Once identified, the goals can be prioritized by the sponsor and champion to develop a scope of the problem. The scope must include a beginning, an end, the included/excluded topics, and the level of detail.

Notional metrics are derived from the goals. Any measurable problem should have a metric to gauge LSS progress. Metrics are dependent on the customer and business requirements. Tollgates are a useful tool for establishing and charting metrics. Tollgates include decision points, reviews, or other opportunities to measure efficiency. More efficiency metrics will be identified in the Measure phase.

### *d. Form and Orient a Team*

The sponsor and the champion have to identify the best candidates for a team. Once the team is selected, it must be introduced to the topic and develop a common understanding of the problem, the project scope, the customer's needs, and the overall process.

The team can then establish ground rules and guidelines in conjunction with a Black Belt and the champion. Rules should ensure that members are open minded, receptive to change, and familiar with the process and subject matter. Tools include a Responsible, Approval, Contributor, and Informed (RACI) chart to establish roles and responsibilities, team-building exercises, and guidance documents (e.g., charter or communications plan).

## e. *Develop Initial Process Maps*

Once the team has become acquainted and shares a common understanding of the problem, its first task is to understand the process as a whole. Process maps are a useful tool in this step. The Supplies, Inputs, Process, Outputs, and Customers (SIPOC) framework provides a guidepost in various steps of LSS. SIPOC is a high-level process map that the team develops to understand and analyze the entire process, including the problem area's scope and impact. Questions during this phase include the following:

- How does this process operate?
- What are the most valuable steps?
- Does the problem area have a significant impact?
- What are the repercussions of addressing/not addressing the problem area within the team's constraints?

The answers to these questions will assist in refining the project and identifying what *will* be covered and what *will not* be covered.

The champion and the sponsor can play a role in this step by providing access to production information and having a firsthand understanding of how the process operates. Other tools for understanding workflow or for use in conjunction with SIPOC include a Pert chart, which captures workflow and output, and value stream mapping (VSM). A value stream map demonstrates the flow of value-added steps to meet product and/or process requirements.

## f. *Create Plans for Overcoming Barriers, Communication, and Schedule*

A thorough understanding of the process enables the team to identify possible roadblocks and solution sets. Often, the greatest roadblocks are barriers to change within an organization. The team is responsible for recognizing any individual or collective assumptions, preconceived notions about the outcome, or reluctance to press ahead with the LSS project. This risk assessment will include projecting the probability of risk that might affect the project, the customer's willingness to take risks, and how risks can be avoided.

Along with a risk assessment, the team should develop a communications plan. This plan will identify the project's purpose and audience and contain a concise message that captures the specific problem area, requirements, schedule, and deliverables. All these planning tools and documents will need to be validated with the sponsor and vetted with the team.

The team must also establish regular meeting times and locations that are convenient for all members, identify a team leader, and initiate communication with the sponsor and the champion.

### *g. Finalize the Problem Statement with a Charter*

The last step in the Define phase is refining the problem and finalizing a problem statement. This process is the culmination of the observations, research, and increasing level of detail obtained in the previous steps. A final problem statement will capture how the problem fits into the bigger system. It can be referenced throughout the project to provide direction and ensure the team stays on track. A clear problem statement will guide future work in process improvement and help to orient a team.

The champion plays a crucial role in this phase by ensuring that the problem clearly addresses the objectives. Effective communication will help to avoid misdirection or mistakes throughout the LSS improvement process. An effective problem statement will include quantifiers and a description of the impact on the entire product or process.

When all of the steps have been completed, a charter* is finalized to capture the problem statement, describe the problem area, identify the project's scope and the specific defect to be addressed, establish a time line, and officially designate the team. Once the entire team understands and agrees to the information in the charter, it can decide to proceed with collecting information and conducting analysis to improve the process and remedy the problem area.

## ***2. Measure Phase***

The Measure phase of LSS includes the development of a data-collection methodology to capture pertinent aspects of the current processes and their outputs, the collection of the data, and the establishment of a baseline for determining improvement. Measure often includes an analysis of the measurement system and process capability. The SIPOC provides a guidepost in various phases of LSS, starting in the Measure phase. The Measure phase is

---

* Some practitioners refer to this as the *statement of work* and reserve the term *charter* for the problem statement from the customer.

not necessarily time or labor intensive. Before the team starts its research, it should look for *quick wins* or solutions that are low risk, readily available, and require minimal analysis. The following subsections describe the activities during the Measure phase.

## a. Develop and Analyze the Process Maps

During the Measure phase, processes from the Define phase will be populated with data. Based on findings from the Define phase, the team will examine the process maps more closely to evaluate the current processes, set data-collection goals, and identify opportunities for improvement based on the data. It will observe and record process steps, including inputs, outputs, reviews, setup activities, reporting requirements, workplace skills, operations, and equipment. Based on these observations, it will identify areas for improvement, which should align with the initial findings from the Define phase.

VSM is the primary tool to identify areas of waste and to ensure that all processes contribute to output and the primary function of the product. Every step must be purposeful and account for what the customer values. VSM identifies places to collect data by breaking down all the steps in the process until waste and variations become evident. Tools including SIPOC, data blocks, and "walking the process" can contribute to the development of a value stream map.

The process maps identify the boundaries of the problem area, as scoped and finalized during the Define phase. A flowchart or other graphic representation can assist in seeing how the process evolves from conception to development and completion, with consideration for crucial decision points. A high-level process map will consider the role of the customer, supplier, and producer, but can also be scoped down to specific processes where a problem may occur, such as shipping or billing. High-level analysis can identify areas of overlap and inefficiency, whereas narrower and detailed maps can signal areas of waste on a smaller scale within one step of the process. Observing the process maps strengthens the team's understanding of the process and lends credibility and insights to the Measure phase.

## b. Prioritize Measurement Tasks

Prioritization is fundamentally based on two interdependent principles: what is most important to the customer and what has the most significant impact on the process. A more thorough prioritization analysis will be conducted in the Analyze phase; however, for the Measure phase, the team must gauge where to focus its data-collection activities, what matters most to the customer, and which steps contribute the highest value to the output. Measuring the process entails observing where the process currently is and identifying the possible ideal state. The latter step requires prioritizing improvement opportunities to direct investments. This process is called *effective utilization*. With the help of the sponsor, the team will be better equipped to understand the process and to identify which tasks are most important for achieving a satisfactory final product.

Prioritization will relate back to the goals established by the customer and team during the Define phase. Sponsor interviews and further analysis of the value stream map can be implemented to answer the question, "Where is the greatest value?"

### *c. Identify Metrics*

Metrics must be customer focused and capture benefits in the areas of highest value. They must also be specific and quantifiable, leaving no room for errors in judgment. Typical metrics include throughput, inventory, expenses, or any quantifiable steps. One particularly useful metric is the time required to perform an activity in relation to the time available.

Once the metrics are identified, the team can evaluate and weigh the metrics according to the impact and importance of the aspect of the process being measured. This evaluation will enable the team to determine which data will be the most useful. The team must also identify constraints to the metrics and data points and how they affect collection activities and analysis.

### *d. Develop a Data-Collection Plan*

After identifying metrics, the team must identify data needs and develop a plan to collect the necessary data. Insights from the process maps highlight data-collection techniques and how to capitalize on the available data. Data needs will account for sources of variation and repeatability and reproducibility.

Before the data is collected, the team should stipulate exactly what it is looking for and know where to find the data and how to collect, measure, and apply this data to the analysis. Data-collection plans should follow the

SMART metrics. Data must be both qualitative and quantitative. Questions asked during this phase should include the following:

- Where is data available on the high-priority areas?
- How can the data be collected, and is it already captured elsewhere (e.g., an annual report)?
- On what aspects of the process will the team focus?

### e. *Collect Data*

The final step of the Measure phase is to apply the methodology and go into the field to collect the data. This step requires cooperation and collaboration with the sponsor to gain access to data sources and to establish the integrity of the samples. Initial sampling done earlier in the Define and Measure phases can provide useful insights into potential challenges and opportunities for the full collection. Once the data is gathered according to the methodology, it is presented to the team for analysis.

## *3. Analyze Phase*

At this point in the DMAIC process, a problem has been defined, and the necessary data has been identified and collected so that the problem can be understood better. The objective of the Analyze phase is to determine the most critical (high-priority) root causes of the problem being addressed (i.e., sources of variation or deficiency).

The Analyze phase is designed to identify and understand root causes from multiple perspectives. To minimize the likelihood of overlooking critical information, an effort is made to identify as many root causes as possible. The most important root causes are determined on the basis of their impact—especially on the customer. A variety of analysis techniques are used on the data associated with these root causes to gain insights into potential corrective actions to mitigate or resolve them. The following subsections describe the activities during the Analyze phase.

### a. *Identify Potential Root Causes*

The root causes, rather than the symptoms, of the problem should be addressed. A root cause is the underlying reason that a problem occurs.

Taking corrective actions on a root cause will prevent the problem from reoccurring. Taking corrective action on a symptom will treat the symptom but not the problem at hand. Also, this effort should not be limited to searching for a single root cause, because an undesirable effect could have multiple root causes. Several techniques can be used to identify potential root causes. Some of the most common are as follows:

- **Brainstorming.** Brainstorming uses open-ended discussion to capture potential drivers causing the problem. The brainstorming process should capture as many ideas as possible. All ideas must be encouraged, and these ideas should not be evaluated or criticized at this time. Weak, impractical, or infeasible ideas will be eliminated later in the process.
- **5 Whys.** Open-ended questions and answers can be informative. Continually asking the question "why" (as many as five times) helps identify more potential root causes.
- **Fishbone Diagram.** A fishbone diagram organizes the potential root causes into categories. A relationship exists between these categories and the brainstorming process. After all brainstorming ideas are collected, developing an affinity diagram will help in defining the major fishbone categories. Using the 5 Whys can break down the categories (or elements within a category) into smaller components. These categories and the smaller components may represent inputs to the overall process, or they may be found on the process maps.

### *b. Prioritize Potential Root Causes*

Common prioritization techniques can be focused strictly on the root causes already identified. They can also be used to supplement the identification effort by looking at a more expansive situation, not just the specific problem at hand. Prioritization is fundamentally based on two interdependent principles: what is most important to the customer and what has the most significant impact on the process. Common prioritization techniques are as follows:

- **Voting.** Simple voting is a first-cut prioritization method and can use a high-, medium-, and low-importance scale.
- **Pareto charts.** Pareto charts show the relative frequency of the factors (potential root causes) that contribute to the problem. Many times, only a few factors will account for the bulk of the problems.

- **XY matrix:** In an XY matrix, the Xs represent the potential root causes and are usually taken from a fishbone diagram. The Ys are the outputs of the process that are important to the customer, and these may be more encompassing than the immediate problem at hand. Stakeholders are asked to numerically rate the effect of each X on each Y and the relative importance of each of the Ys. In that way, using quality function deployment techniques, the Xs can be prioritized.
- **Failure modes and effects analysis (FMEA).** A FMEA is a disciplined procedure that identifies ways in which a process or a product can fail (failure modes) to meet customer requirements, the reasons why the failure occurred (root causes), and the impact of the failure (failure effects). A FMEA can be conducted on just the steps that affect the problem at hand, the steps in the process encompassing the highest-priority Xs, every step in the process, or anywhere in between. When used for a product, FMEAs can be employed at the system or subsystem level in the early design stage so the design can address the failure modes observed. FMEAs can also be used to analyze new process designs or to improve operational processes. Prioritization is accomplished by considering the degree of severity of the failure, the likelihood of occurrence (taking into consideration the current controls in effect), and the ability to detect the failure mode.

### *c. Analyze the High-Priority Root Causes*

The high-priority root causes are analyzed to obtain greater insight on what to do about them. This analysis lays the groundwork for determining how to improve the situation in the next DMAIC phase. Root cause analysis techniques vary as a function of the level of knowledge of the situation and the availability of data to support that knowledge. The techniques can be simple or complex. At a basic level, a statistical analysis of the overall process or a part of the process can show the relative magnitude of the problem and provide a measure of process performance over time. More complex statistical analyses can be used to understand variations in much greater depth and to predict the outcome of changes. Some common analysis tools used to establish cause-and-effect relationships between the Xs and Ys include the following:

- **Run charts.** Run charts plot the cycle time of different observations. Outliers can then be examined to determine what they have in common.

- **Graphical analysis.** Graphical analysis is used to understand the distribution of the data so that more sophisticated statistical techniques can be applied.
- **Goodness-of-fit tests.** Goodness-of-fit statistics are used to determine how well the data can be characterized by a specific probability distribution.
- **Hypothesis testing.** Hypothesis tests are statistical techniques for comparing properties or determining whether relationships exist among different populations of data.
- **Scatter diagrams.** Scatter diagrams can assist in confirming relationships among causes and effects. They graphically depict something that can be tested by simple linear regression.
- **Regression testing.** Regression tests quantify the relationship (correlation) between input (independent) and output (dependent) variables. Correlations can also be developed among root causes. Correlation, however, does not determine causation.
- **Analysis of variance (ANOVA).** ANOVA gives a statistical test of whether the means of several groups are all equal so that the effects of various treatments can be compared.
- **Quality Function Deployment (QFD).** In building the XY matrix, relationships were developed between root causes and the characteristics important to the customer. QFD extends that concept by examining how specific features, attributes, and/or metrics contribute to what the customer wants. QFD also includes the determination of targets for customer needs and the features, attributes, and/or metrics.

## 4. Improve Phase

The objective of the Improve phase is to determine the actions necessary to change the process and improve performance. Improvements occur through increasing value to the customer and eliminating waste. Improvements are quantified by comparisons to the product- and process-related baselines established in the Measure phase.

Developing potential solutions involves a complex set of activities that should be tailored to the specific situation being addressed. Multiple solution-generation techniques are often employed since different approaches can generate more effective ideas. For example, brainstorming for solutions to the various root causes and the application of Lean principles to identify and eliminate waste are complementary approaches. If brainstorming is used to identify mitigation actions for the root causes, the entire new process could then be made Lean and safeguarded to eliminate mistakes and avoid *back sliding*. The Improve phase ends with determining the most effective mitigation actions and developing a plan to implement them. The following subsections describe the activities during the Improve phase.

### a. Brainstorm Improvement Ideas

Solutions should be generated for all of the high-priority root causes in the XY matrix, one at a time. As many solutions as possible should be generated since the best ideas will be determined later. Team-based brainstorming is a structured and effective way of generating many ideas in a short period of time. The key to successful brainstorming is to keep the creative process going by not putting any limits on the ideas suggested and not evaluating ideas during the brainstorming process.

The team must overcome conventional assumptions and self-imposed constraints. One useful exercise is to tear apart the existing process and challenge everything that it does. "We've always done it this way" is not a reason to continue the same practices in the future. Once some team members suggest new ways of doing things, others become inspired to build on these new ideas.

Brainstorming is typically carried out in a "round-robin" fashion. When the flow of new ideas slows, the team should begin the process again. When these iterations have finally ended, the team should review all ideas to ensure that everyone has a common understanding of their meaning.

Completion of the review should be followed by an initial screening of ideas. Some ideas may violate a law or be too risky. All impractical or infeasible ideas should be eliminated at this point. All remaining ideas will imply changes to the process and will form a set of potential to-be processes.

### b. Generate Solutions to Make the Process Leaner

Some basic Lean principles are as follows:

- Specify what creates value from the customer's perspective.
- Identify the steps in the process chain.

- Implement changes needed to improve process flow.
- Produce only those things that are demanded by the customer.
- Continuously remove waste from the processes.

Waste must be identified before it can be eliminated. The seven areas of waste are as follows:

- Rework/correction
- Over production
- Unnecessary processing
- Excess conveyance/transportation and inventory
- Unnecessary movement
- Waiting
- Unnecessary investment

Process variation clearly results in waste. The root causes of process variation determined in the Analyze phase represent sources of waste. Therefore, to identify waste, special attention should be paid to those steps in the process associated with the root causes.

Waste can often be identified by examining the overall process as it works today and as it might work in the future given the implementation of some of the ideas developed during the brainstorming process. Determining the time required for each step in the process and then identifying the value added by that step provide an opportunity to identify waste. If a bottleneck in a process takes time but does not add any value to the customer (and is not a mandated requirement), it is waste. Waste is identified throughout the entire process, not just the areas where root causes of variation were found.

These efforts, when completed, accomplish two important objectives:

- Identify potential waste associated with the changes proposed to fix the high-priority root causes.
- Identify waste in other steps in the process. While these areas of waste may not be associated with the problem at hand, they undoubtedly add cost to the process and may represent some "low-hanging fruit" in terms of improvement to the process.

### c. Determine Ways to Eliminate Waste

All waste should be eliminated; however, the elimination process depends on the situation and is not always obvious. For example, if the process owner is involved, some step in the process can be taken away without any additional effort (e.g., transportation waste can be eliminated by moving things closer together). On the other hand, taking a step away can create transition issues that result in other changes when the elimination occurs. For example, excess inventory levels can only be eliminated by changing many elements of a process.

The FMEA begun in the Analyze phase contributes to waste elimination. The initial FMEA identified the failure mode, failure effect, and cause of failure. All failures represent defects or waste. By identifying the controls needed to eliminate the cause of failure, recommended actions can be developed to mitigate the defect. If new processes are being created, augmenting the initial FMEA may be necessary since any process change may result in unintended consequences.

Another useful waste elimination technique is the Theory of Constraints (TOC). A constraint is anything that impedes throughput. The binding constraint in a process is the step that wastes the most time. The TOC seeks ways to mitigate this situation so that some other step in the process becomes the binding constraint.

Other common waste elimination strategies include the following:

- **Just-in-time (JIT).** Inputs to each step in the process arrive only when needed and in the correct quantity to reduce inventory holding costs.
- **Pull production systems.** Such systems are essential in a JIT environment. The supply of the input to any step in the process is triggered by a signal. Purchase pull systems and replenishment pull systems reduce waiting time.
- **Parallel processing.** This strategy allows independent steps in the process to proceed simultaneously, not serially.
- **Process balancing.** This strategy attempts to ensure optimal use of people, floor space, capital assets, and material. In a parallel processing environment, process balancing will equalize the time needed to accomplish parallel tasks (although the total effort may be different).
- **Process flow improvement.** This strategy reduces the time needed to complete a step in the process through simplification accomplished by an improved layout and/or standardized operating procedures.

In some situations, the process of eliminating waste may not be clear. Using a design of experiments (DOE) is a way to better understand the real world. Under DOE, deliberate and systematic changes of input variables are made to observe the corresponding changes in output variables. This approach is often a cost-effective method of determining whether specific actions will work as expected before they are actually implemented.

### *d. Determine Ways to Sustain Waste Elimination*

Application of the 5S standards is one way of making gains sustainable. 5S is a process for creating and maintaining an organized workplace. The 5Ss are as follows:

- **Sort** through and remove clutter and unneeded items.
- **Set** the workplace in order and make it obvious where things belong.
- **Shine** the workplace from top to bottom while identifying hazards and mechanical problems.
- **Standardize** guidelines for the 5S conditions.
- **Sustain** the gains by making an organized workplace part of the culture and the daily routine.

The visual workplace is a process management technique that complements the 5Ss. Visual controls communicate necessary information clearly and quickly. They highlight exactly what is and what is not efficient or effective. In that way, corrections can be targeted to the specific problem.

Standardizing work and mistake-proofing are other ways to sustain the improvements. Mistakes add cost and provide no value to the customer. Mistake-proofing involves using wisdom and ingenuity to create devices that allow the job to be done 100% defect- or error-free 100% of the time. The devices can be designed for prevention, detection, warning, or self-correcting control and can simplify the job requirements. For example, something can be made tamperproof if that is a problem that needs to be corrected.

### *e. Determine the Most Important Actions to Be Taken*

In most circumstances, trying to accomplish everything at once is too difficult. Therefore, payoffs from the potential actions should be explored so that the best set of recommended activities can be pursued in the right order. Different ranking tools can be used to help prioritize these activities. For example, a benefit–effort matrix can be created, and emphasis can be placed on low-effort actions, especially those with a high payoff. Ranking criteria can also be developed, and each action can be evaluated to determine those with the highest priority. The team should not lose sight of the most important root causes identified in the Analyze phase, because addressing these issues will often provide the most effective near-term course of action.

Another important consideration is complexity. Even when a customer considers complexity to be value added, actions that increase complexity should be carefully reviewed before they are incorporated into an implementation plan. Complexity drives costs and increases the potential for error. In some cases, the elimination of complexity is the most cost-effective approach.

Determining the most important actions to be taken always involves unknowns. Solutions may not work as expected. A pilot program can be a valuable step for testing solutions on a small scale before the entire process is affected. It provides an opportunity to discover and mitigate problems earlier in the process. It also provides an opportunity to increase buy-in on the final solution.

### *f. Create an Implementation Plan*

Changes cannot be made until a plan is devised to execute them. The implementation plan includes what needs to be accomplished and the actions needed to do it. Large tasks should be divided into subtasks to make them more manageable. The plan should detail the time and resources required to do the job, the expected start and completion times for the key actions, and the people, equipment, supplies, and money needed to accomplish each action.

An effective implementation plan also describes the roles of all of the key stakeholders involved. Stakeholders include the people who:

- are responsible for making the change,
- contribute to the nature of the change,
- must be kept informed of the change, and
- approve the change.

Implementation plans should address the effects of making the change. For example, when a process becomes more efficient, fewer people will be needed. The implementation plan should show how people will be redeployed into other productive work.

## *5. Control Phase*

The objective of the Control phase is to ensure that the implementation plan developed in the Improve phase achieves the desired effect. This objective is achieved by establishing controls to help manage changes. Controls include training, communication, and an implementation monitoring effort. Since few plans are executed as expected, the final part of the Control phase is a corrective action process.

Another element of the Control phase, which is not discussed below, includes LSS project wrap-up efforts. The results of the work should be publicized, and team members should be recognized for their contributions. Opportunities to deploy similar changes elsewhere in the organization should be sought. Finally, lessons learned should be systematically captured. The following subsections describe the activities during the Control phase.

### *a. Execute the Implementation Plan*

The first step in the Control phase is to execute the implementation plan created in the Improve phase.* No matter how much effort goes into preparing an implementation plan, its execution rarely proceeds as expected—even when a pilot implementation is used.

Changing the status quo always generates opposition. Cultural adjustments and training are necessary. In some cases, the implemented changes do not produce the predicted effect. Consequently, the implementation effort must be controlled. Without this monitoring, sustaining the gains may not be possible.

### *b. Establish Controls*

Implementation control is accomplished with a control plan. A control plan identifies the actions that are required at each phase of the process to ensure that all process outputs will be in a state of control. It tracks all of the inputs to each phase of the process, describes how the inputs and

* Some LSS practitioners execute the implementation plan in the Improve phase.

outputs are being measured, monitored, and controlled, and states what should be done when something is not in control.

Planned periodic status reviews should be conducted to monitor implementation progress. The control plan identifies metrics that measure the extent to which the implementation plan is being executed as designed, is having an impact at a subprocess level (e.g., errors are being reduced), and is providing the customer with greater value (i.e., the overall process has been improved). The execution-related metrics are often process related and refer to inputs more than to outputs. At the subprocess level, the metrics become output related, and when dealing with the customer, the metrics usually represent outcomes. Since the old measurement system may not be adequate for the new process in some instances, the measurement system should be evaluated to determine whether it meets the new requirements.

The control plan also identifies the data that should be collected or audits that should be conducted to obtain the desired progress measurements. The auditors should be unbiased and qualified, and the data collected should be measured against a defined standard.

Another key aspect of the control plan is change management. Change takes time and always causes resistance. A control plan attempts to reduce that resistance by ensuring that everyone is informed about the changes and trained to effect the changes. Beyond conventional delivery mechanisms, coaching and on-the-job training are also effective and should not be overlooked. New work instructions, policies, and standard operating procedures must be clear, in place, and communicated thoroughly to all of the stakeholders.

### c. *Take Corrective Actions (as Appropriate)*

The established controls form a feedback system to report deviations in actual outputs from desired levels. Actions need to be taken when the implementation plan is not producing the predicted effect. These actions must be timely and effective and should prevent the recurrence of the problem. The underlying cause of the problem must be identified, and the FMEA may have to be revisited.

A corrective action plan documents the specific actions that need to be taken. Just like the implementation plan, it must identify the required steps and resources, the time line, and the roles of all of the key stakeholders. The specific actions should be determined in accordance with a corrective action process that establishes what actions are needed as a function of what problems are encountered. For example, implementation may be too slow, it may not have fixed the problem, or it may have attacked the wrong problem. The corrective action process also tracks who is responsible and whether the corrective actions have been successfully accomplished.

## B. The Design for Six Sigma (DFSS) Methodology

### *1. The Relationship between DFSS and LSS*[*]

DFSS is an approach for applying the LSS thought processes (especially the Six Sigma aspects) to the design of *new* products and processes. LSS and its five-phase DMAIC methodology are not strongly focused on design but on adding value for the customer and supplier and reducing waste or variation in existing products and processes. Six Sigma's origins are derived from the manufacturing floor and, consequently, it is commonly used at the operational stage in the product/process life cycle. At this point in the life cycle, defects (variations) are typically easy to identify but costly to fix.

Preventing defects during the initial design phase produces far greater leverage. At this stage, defects are relatively easy to fix but are more difficult to observe or predict because the design elements have not been finalized. Once the design is complete, most of the costs are locked in. Therefore, the greater the effort in preventing defects (variability) during design, the larger the payoff that can be realized later. The problem with using LSS is that it would attempt to improve something that does not exist—which is why DFSS was created. It adapts LSS to this new product/process situation.

Applying LSS on an existing product/process reduces variability to fewer than 3.4 defects per million units (the Six Sigma level of performance). Figure 3.2 depicts the application of LSS in reducing variation. The curve labeled "Traditional Product" is meant to depict a situation for a typical defense product. The x-axis represents the performance of some important critical-to-quality (CTQ) attribute of the system. As far as the customer is

[*] The material in this section was developed from Gene Wiggs, "Design for Six Sigma Introduction," a General Electric (GE) Aviation briefing (September 2005) and Martha Gardner and Gene Wiggs, "Design for Six Sigma: The First 10 Years," vol. 5, *Proceedings of GT2007 ASME Turbo Expo 2007: Power for Land, Sea, and Air* (Montreal, Canada, May 14–17, 2007).

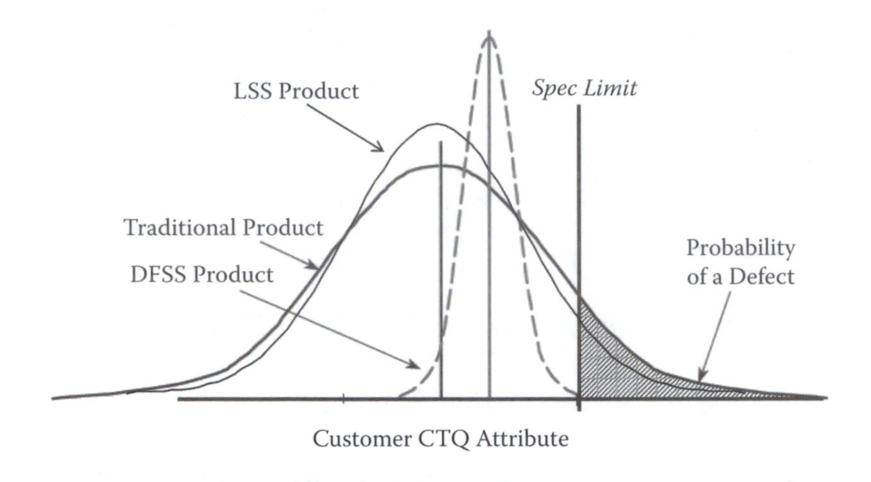


**Figure 3.2 Customer CTQ attribute. (Reproduced with permission from material provided by Gene Wiggs of the General Electric Company [GE].)**

concerned, the mean should be as high as possible, as long as the number of defects (indicated by the shaded area) is not too large.

The application of LSS principles would improve the process and reduce the defect rate. In Figure 3.2, the mean performance value would be the same as the Traditional Product curve; however, the distribution would be tighter in a way that experience has shown would reduce the number of defects by about one third.

As shown in the curve labeled "DFSS Product" in Figure 3.2, the use of DFSS principles will tighten the variability about the mean even further by quantifying the risk and enabling variation levels to be much smaller than with other design philosophies. This enhanced reduction in variation implies much less than 3.4 defects per million units (i.e., a seven, eight, or nine sigma level of performance). Furthermore, it allows the possibility of a specification change by improving the mean performance while maintaining a Six Sigma (or some other acceptable risk) level of defects.*

Such a specification change provides significant potential for performance improvements, competitive advantage, and cost savings. For example, suppose a design change that significantly reduces vibration can be made to an aircraft. This change enables the aircraft manufacturer to reduce insulation and thereby reduce weight. The aircraft manufacturer could then change the

* Figure 3.2 is drawn to illustrate a point. Technically, the height of the DFSS curve should be far greater so that the area under the DFSS curve is the same as the area under each of the other two curves.

vibration specification on the engine because of the improved performance, using DFSS to give a significant advantage to the engine manufacturer.

DFSS can therefore be defined as a systematic methodology for (1) using tools (e.g., QFD, reliability modeling, scorecards, design and analysis of computer experiments, Monte Carlo simulation, accelerated life tests, FMEAs, optimization), (2) training on the steps in the process and the use of the tools, and (3) employing conventional Six Sigma measurement system analysis to develop products or processes that meet CTQ customer expectations by managing variation in design. To achieve desired performance over time, DFSS matches designs with the manufacturing capabilities (design for producibility), the operating environment (design for reliability and robust design), and the costs (design for affordability). It enables the design to be manufactured so that all customer and regulatory technical requirements are met with minimal defects.

## 2. How DFSS Affects the Design Process

Instead of DMAIC, DFSS uses the following five-phase process—Define, Measure, Analyze, Design (and Optimize), and Verify (DMADV)—to *design* quality in, rather than to *test* quality in.*

The Define phase establishes the requirements. The CTQ customer expectations flow down to subsystems and components, and quality goals are set. One type of quality goal is the probability of a defect. This goal is closely related to reliability; however, defects can also result from mistakes in the field and/or errors in the design. The other type of quality goal is more operationally oriented. It deals with the probability of meeting mission performance needs and therefore is an indication of robustness.

The measurement systems are determined in the Measure phase. These systems are used to collect and understand data on the actual production variation of CTQ elements of existing products/processes overall (e.g., weight)

* This approach to DFSS is not universally accepted. Different companies have different implementations. The names are different. The acronyms are different. The number of steps is different. However, the activities are similar. See, for example, Lisa A. Reagan and Mark J. Kiemele, *Design for Six Sigma—The Tool Guide for Practitioners* (Bainbridge Island, WA: CTQ Media, 2008). This book uses identify, design, optimize, and verify (IDOV) as its framework. The implications for synergies with Value Engineering (VE) are the same. Because of the absence of a standardized approach to DFSS, a description of the steps analogous to what was done for the job plan or DMAIC has not been constructed. A reference for the DMADV approach is Eric Maass and Patricia D. McNair, *Applying Design for Six Sigma to Software and Hardware Systems* (Upper Saddle River, NJ: Prentice Hall, 2009).

and on the actual producibility, reliability, and cost of their individual parts to create a baseline. The variation change and specification shift depicted in Figure 3.2 would result from design changes made on the existing products/processes to create the new products/processes or the use of completely new technologies. In the latter case, baseline levels would not be available.

Conceptual designs are developed in the Analyze phase. These conceptual designs enable architectural decisions. An architectural decision involves something that affects several CTQs simultaneously. For example, in the design of a new car, determining whether that car will have a six-cylinder engine option or a four-cylinder engine option is an example of an architectural decision based on conceptual design. If a four-cylinder engine could satisfy all customer expectations for acceleration and performance, several additional CTQ customer requirements (e.g., fuel economy, vibration, noise, space, weight, crash protection, and fuel-based emissions) are affected.

Also, the framework for making risk assessments is established in the Analyze phase. Unlike DMAIC, which is concerned with statistics (what is actually happening), DMADV deals with probabilities (what may occur). These probabilities can be derived from computer simulations based on system physics or from estimates based on historical data. Consequently, risks associated with quality goals can be quantified on quality scorecards for CTQ requirements at the system level and for cost, reliability, and producibility at the part level. These scorecards show target values, expected values, standard deviations, and defects per million units. This defect rate is used to make quantitative, risk-based design trades. It quantifies how much risk is acceptable and therefore drives design decisions.

System and component models (both physical simulations such as designed experiments and physics-based models) are constructed in the Design (and Optimize) phase. These models are used to predict the expected performance of CTQ requirements at the system level and reliability and producibility at the part level. The models are used to evaluate how the variation of potential new part designs will differ from the baseline data collected in the Measure phase. The use of mathematical transfer functions enables an estimation of how predicted changes in mean and variance at the part level will affect the variation of CTQ attributes for the system. With an understanding of the process capability of the production line, a determination is made of whether designs with sufficiently small variation are producible. These model-developed expected values and variation are used to complete the scorecards and calculate expected defects per million to determine the acceptability of the risk. Consequently, critical design parameters can be controlled analytically so that fewer prototypes have to be built and tested, even though building prototypes to verify key theoretical findings from the models is common. In addition, design revisions are minimized. This approach is the beginning of design for producibility and affordability.

The optimization element in the Design (and Optimize) phase focuses on reliability, robustness, and tolerance. The goal is to optimize a detailed design that minimizes cost by creating robust, producible designs that are tolerant of variation at acceptable risk levels associated with reliability and performance. Figure 3.3 illustrates design for reliability. It shows the probability distributions of the force that the part can withstand (strength) and the force that the part may be subject to (stress). The overlap of the stress (that may be encountered in the field) and strength (the design characteristic of a part) curves indicates places where a part can fail. Designs mitigate this by moving the stress curve to the left or the strength curve to the right.

Figure 3.4 illustrates an application of robust design and shows the importance of determining how much the design has to back off from the deterministic optimum to account for variation in the independent variable. In the figure, the optimum value of $X$, the independent variable representing something that might happen in the field, is where the regression curve achieves the target $Y$ value for some CTQ attribute, or approximately $X = 6$. However, $X$ is subject to variation with a standard deviation of 2. Therefore, a value of $X$ three standard deviations above 6 would produce a $Y$ value above the upper limit. For an $X$ of approximately 2, three standard deviations in either direction yield a $Y$ value between the bounds.

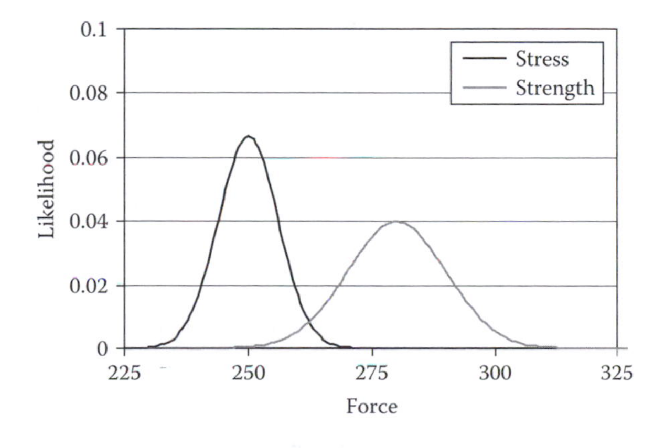


**Figure 3.3 Stress and strength in design. (Reproduced with permission from material provided by Gene Wiggs of the General Electric Company [GE].)**

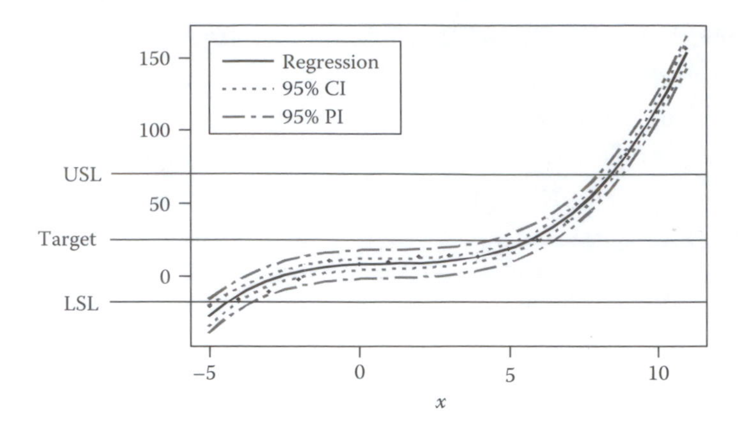


**Figure 3.4 Application of robust design. (Reproduced with permission from material provided by Gene Wiggs of the General Electric Company [GE].)**

The modeling, design development, scorecard completion, risk evaluation, and design optimization discussed previously are performed iteratively in this phase. These steps are performed for system architectures in conceptual designs, for second-order architectural configurations in preliminary design, and for design-to-print package development in detailed design.

In the Verify phase, design verification using pilot, preproduction, and production units ensures that all requirements are met. DMAIC-like control plans are developed and then the transition to production occurs.

## 3. Complexity

Complexity, as measured by the number and difficulty of critical elements or actions involved, influences the likelihood of defects resulting from both variation and mistakes.* Therefore, effective design encompasses more than the control of variation. Many failures in the field are associated with errors. These errors can be mistakes by the user (e.g., inadequate human factors engineering), by the maintainer (e.g., wrong part replaced because several parts look similar), by the manufacturer (e.g., the product was assembled incorrectly because of apparent symmetries), or by the design team.

For example, consider a design that requires 80 holes to be drilled to fasten two parts together. Serious consequences can occur if only 79 holes are drilled. The fact that the wrong number of holes was drilled is a process control problem, *not* a statistical variation problem. Another example would be

* See C. Martin Hinckley, *Make No Mistake* (New York, NY: Productivity Press, 2001).

two similar parts that an assembler could confuse and put in the wrong place. A well-thought-out design for manufacturing and assembly, coupled with a set of best process control practices, should prevent many of these mistakes.

The risk associated with these errors cannot be quantified by the physics-based modeling approach described previously or by process variation distributions. An FMEA can be created in the Analyze phase and used in the Design phase to identify the risks and the measures taken to mitigate them. For example, design processes can be error proofed to some extent by incorporating automation, checklists, and nonadvocate reviews. The team should work in the Design phase to reduce complexity by using standard processes and standard parts and features and by decreasing the sheer number of parts.

# Chapter 4

# Comparison of VE and LSS Methodologies

## A. VE and LSS Cross-Reference

Figure 4.1 overlays the Value Engineering (VE) and Lean Six Sigma (LSS) activities portrayed in Figure 2.1 and Figure 3.1. The steps in our discussion of VE and LSS represent a synthesis of information from literature and training material to provide the reader with an appreciation for the logical flow of events that transition smoothly from one activity to another, working toward a solution.

Figure 4.1 identifies correlations, similarities, differences, and opportunities for synergies between the two methodologies. Gray boxes represent the VE job plan, and white boxes represent the Define, Measure, Analyze, Improve, and Control (DMAIC) methodology. In some places, a one-to-one correlation exists. For example, the single step in the VE Orientation phase to Establish Evaluation Factors matches Identify Metrics in the DMAIC Measure phase. In other places, the mapping is more complex. Cases where multiple DMAIC activities map into a single VE job plan activity are striped. The activities represented by diagonal stripes identify structural differences (in the Define, Measure, and Control phases), and the activities in dark gray shading represent a difference in the analytical approach (in the Analyze and Improve phases). The corresponding number beside these boxes identifies the number of overlapping DMAIC activities. These areas represent the

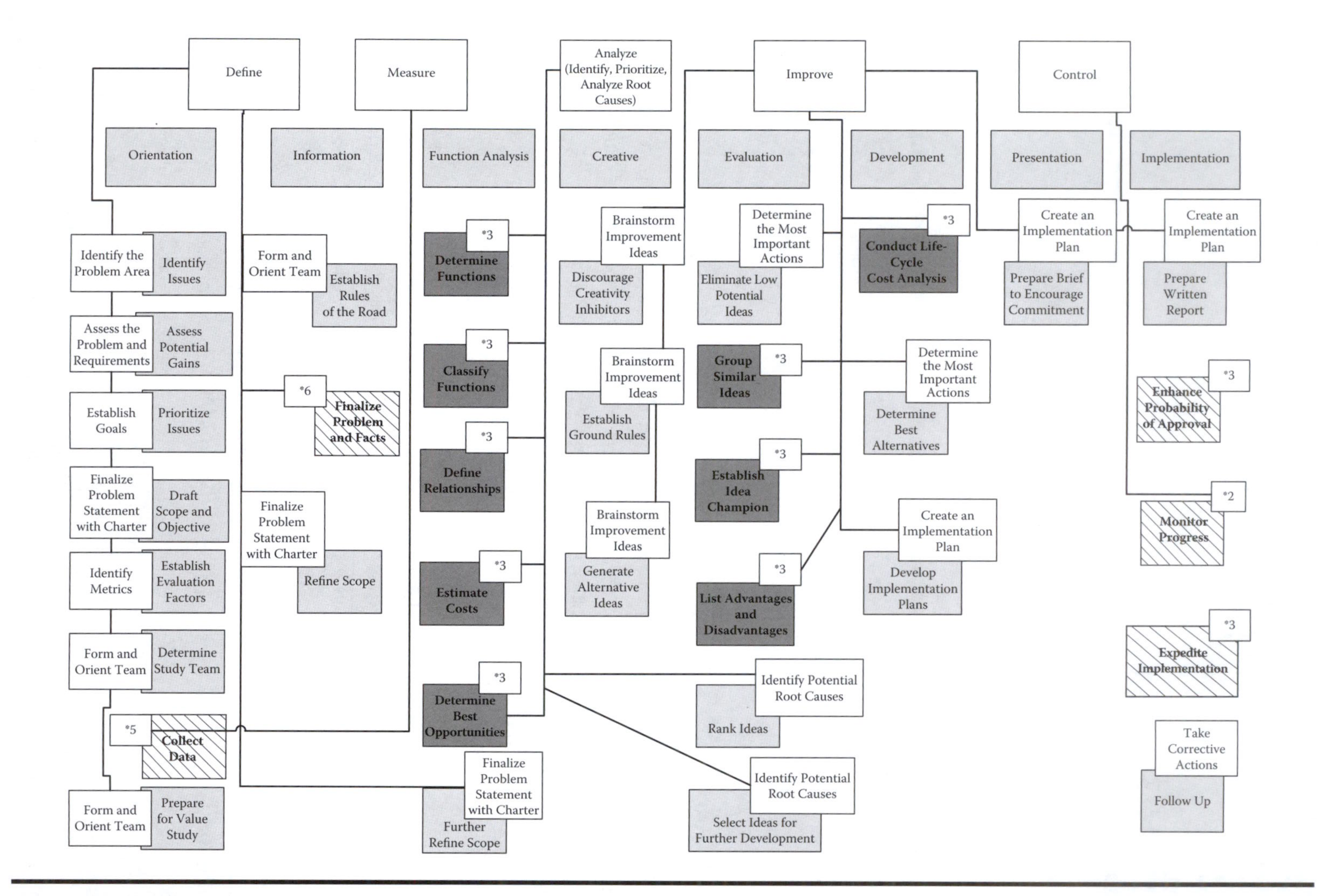


**Figure 4.1 VE job plan cross-referenced with DMAIC methodology.**

highest potential for synergy. As an example of how to navigate this chart, consider the Determine Functions activity in the Function Analysis phase of the VE job plan. This activity corresponds to a DMAIC box with *3, and is linked to the Analyze phase. This means that the Determine Functions activity correlates to three activities in the Analyze phase of the DMAIC methodology. Table 4.1 shows a breakdown of these numbered boxes. Appendix A contains more detailed cross-reference charts.

**Table 4.1 Notes for Figure 4.1**

| *Structural (Define, Measure, and Control Phases)* |
|---|
| **Collect Data (5):** Develop initial process maps, develop and analyze the process maps, prioritize measurement tasks, develop a data-collection plan, collect data<br>**Finalize Problem and Facts (6):** Develop initial process maps, develop and analyze the process maps, prioritize measurement tasks, develop a data-collection plan, collect data, finalize problem statement with a charter<br>**Enhance Probability of Approval (3):** Create plans for overcoming barriers, communication, and schedule; execute the implementation plan; establish controls<br>**Monitor Progress (2):** Execute implementation plan, establish controls<br>**Expedite Implementation (3):** Execute the implementation plan, establish controls, take corrective actions (as appropriate) |
| *Analytical (Analyze and Improve Phases)* |
| **Determine Functions (3):** Identify potential root causes, prioritize potential root causes, analyze the high-priority root causes<br>**Classify Function (3):** Identify potential root causes, prioritize potential root causes, analyze the high-priority root causes<br>**Define Relationships (3):** Identify potential root causes, prioritize potential root causes, analyze the high-priority root causes<br>**Estimate Costs (3):** Identify potential root causes, prioritize potential root causes, analyze the high priority root causes<br>**Determine Best Opportunities (3):** Identify potential root causes, prioritize potential root causes, analyze the high-priority root causes<br>**Group Similar Ideas (3):** Generate solutions to make the process leaner, determine ways to eliminate waste, determine ways to sustain waste elimination<br>**Establish Idea Champion (3):** Generate solutions to make the process leaner, determine ways to eliminate waste, determine ways to sustain waste elimination<br>**List Advantages and Disadvantages (3):** Generate solutions to make the process leaner, determine ways to eliminate waste, determine ways to sustain waste elimination<br>**Conduct Life-Cycle Cost Analysis (3):** Generate solutions to make the process leaner, determine ways to eliminate waste, determine ways to sustain waste elimination |

While Figure 4.1 is useful for comparing the two methodologies, its primary purpose is to identify areas that have the greatest opportunities for synergy. Stipulations and limitations of the cross-reference include the following:

- Not every study executes every step explicitly.
- The order of the steps may differ in practice.
- The steps can be performed recursively.
- The rigor of analyses varies in practice.

The tools used for analysis may be different. LSS uses many more statistical analysis tools while Design for Six Sigma (DFSS) tools have a probabilistic focus. VE's toolset is more mechanistic in nature. Appendix B lists commonly used tools for the three approaches.

Consequently, the cross-referencing shown in Figure 4.1 is not absolute. Nevertheless, the mapping does demonstrate some important points. In some areas, the methodologies

- are similar,
- have different levels of detail, and
- take a different perspective.

The differences do not imply that one methodology is better than the other, nor do they imply weaknesses. Instead, they indicate opportunities where both approaches can be used together to achieve better results.

## B. How VE Can Benefit from LSS

When LSS establishes goals in the Define phase, customer communication tools such as Likert scales, surveys, interviews, and focus groups are used. The VE counterpart, *Prioritize Issues*, is more focused on potential gains and the feasibility of implementation. More formalized customer communication would help with decision maker acceptance and approval of VE-generated recommendations.

LSS has a more detailed front-end process for data collection. Whereas the VE methodology simply states that the data should be collected, LSS creates and analyzes process maps, determines and prioritizes measurement systems, and establishes a formal data-collection plan.* When VE finalizes the problem and facts in its Information phase, it often uses Quality Function Deployment (QFD) to obtain a better understanding of the data and data sources in the context of the problem. The LSS Supplies, Inputs, Process, Outputs, and Customers (SIPOC) framework is used to understand the entire process and where the problem fits in. VE's use of SIPOC could add insight to its Function Analysis phase.

LSS also has a more disciplined approach toward implementation. VE simply creates an implementation plan and follows typical best practices to execute it. The LSS control plan is a formal activity to ensure that execution proceeds as planned, with specific metrics identified in advance. Furthermore, LSS includes a formal corrective action plan (sometimes as a separate process), which is not an unambiguous part of the VE methodology.

These differences represent areas in which incorporating some LSS features would likely improve the VE methodology. These synergies would help formalize the VE process to reduce the likelihood of overlooking important information needed to help determine a course of action. They would also improve the likelihood of successful implementation.

A VE continuous process improvement project, conducted on August 3, 2009, by the Defense Supply Center Columbus (DSCC), provides an example of how the more structured LSS data-collection process can help improve the VE methodology. DSCC uses VE to identify alternative lower-cost suppliers for the items it manages for the armed services. The process is data intensive. First, research is conducted to determine candidate items with a high potential for cost reduction through the development of new sources. Second, functional information on the item (e.g., drawings, specifications, stock samples, special markings) is collected to determine whether potential new sources are qualified and interested in bidding.

The data-collection process used by DSCC VE analysts was not well defined. Data sources were not standardized. Electronic data-collection tools were not consistently used. Therefore, the LSS project objective was to increase the likelihood of successfully qualifying alternative sources by improving the data selection tools, sources, and collection processes for identifying candidates. Activities during the LSS project phases were as follows:

- **Define.** The problem statement and business case for action were developed. The team was formed, and the expectations were established.

* The data collected is different. VE focuses on cost data. LSS focuses on process performance.

- **Measure.** The process value stream map was created, and metrics were defined.
- **Analyze.** Issues with the *as is* process were listed, along with associated explanatory comments, suggested improvements, and expected results for implementing the suggested improvements.
- **Improve.** Recommendations were sorted by payoff potential and ease of implementation. A prioritized list was developed, and an implementation plan was drafted.
- **Control.** Plans to review the effects of recommended changes were established to ensure that they were being implemented and were achieving the expected results.

These activities transformed an ill-defined VE data-collection process into one that follows a standardized and robust data-collection process formulated by developing and analyzing process maps to prioritize measurement tasks. Similar benefits can be achieved in other VE applications.

## C. How LSS Can Benefit from VE

VE and LSS develop solutions to problems from different perspectives. Some of the most important distinctions are as follows:

- VE explicitly considers cost by collecting cost data and using cost models to make estimates for all functions over the life cycle. LSS reduces cost by eliminating waste and reducing variation through the use of statistical tools on process performance data. Exclusive emphasis on waste can be contradictory to reducing life-cycle cost. In VE, some waste can be tolerated if it is necessary to achieve a function that reduces the life-cycle cost. Safety stock to mitigate occasional supply disruption is a good example.
- In determining what should be changed, VE's function analysis identifies areas that cost more than they are worth, while LSS identifies root causes of problems or variations. VE's separation of function from implementation forces engineers to understand and deliver the requirements.
- For required functions that cost more than they are worth, VE uses structured brainstorming to determine alternative ways of performing them. LSS brainstorms to identify how to fix the root causes. Because functional thinking is not the common way of examining products or

processes, VE augments the structured innovation process in a way that generates a large number of ideas. Shingo[*] suggests that VE is one of the most effective techniques for attacking the fundamentals of a problem. Enormous improvements are possible by determining which functions are really required and then determining how to best achieve them.

- VE develops solutions by evaluating the feasibility and effectiveness of the alternatives. LSS emphasizes solutions that eliminate waste and variation and sustain the achieved gains. VE eliminates waste in a different way. VE separates the costs required for basic function performance from those incurred for secondary functions to eliminate as many non-value-added secondary functions as possible, improve the value of the remaining ones, and still meet the customer requirements.
- An LSS focus on quick wins may preclude an in-depth analysis of the situation. Without analysis, projects can suboptimize or even work in opposition to one another. Using function analysis should prevent this suboptimization.

A systems engineering approach to problem solving involves analyzing the system as a whole. The major components should be understood individually and collectively in an operational construct. Then, the system can be decomposed to understand proposed changes using function analysis.[†] Problems can occur if this approach is not followed, as the following real example illustrates.[‡]

The Navy's Standard Missile is a surface-to-air defense weapon. Its primary mission is fleet area air defense and ship self-defense. It also has a secondary antisurface ship mission. It uses a mirror for celestial navigation. At one point in time, the supplier of the one-piece mirror being used announced the phase-out of the current product. The one-piece mirror was replaced with a three-piece mirror that reduced the unit price, provided the same performance, and was readily available.

Unfortunately the three-piece mirror had different mass properties than the one-piece mirror it replaced. The firmware for the guidance section had built-in compensation factors for the mirror's mass properties, and with the new mirror, it was compensating for the wrong measurements. This

---

[*] Shigeo Shingo, *Study of the Toyota Production System from Industrial Engineering Viewpoint* (Tokyo, Japan: Japanese Management Association, 1981).

[†] A systems-level FMEA also provides useful insight.

[‡] James R. Vickers and Karen J. Gawron, "A Systems Engineering Approach to Value Engineering Change Proposals," paper presented the 2009 DMSMS Standardization Conference (Orlando, FL, September 21–24, 2009).

problem led to a costly shutdown of the production line until the problem was fixed.

Since certain problems can be more readily, effectively, or thoroughly managed by using LSS, VE, or both, the full range of options for solving the problems should be explored. The next chapter summarizes what the recent VE literature says on opportunities for collaboration and the benefits of integration.

# Chapter 5

# Opportunities for Synergy

## A. Literature Review

Value Engineering (VE) and Lean Six Sigma (LSS) have limitations, but synergizing the two methodologies optimizes similarities and provides the potential to overcome their unique limitations. The literature on LSS and VE[*] analyzes the strengths and weaknesses of the methodologies and highlights opportunities for collaboration. The literature examining these methodologies points to two primary areas where VE can contribute: scope and creative tools such as the FAST diagram. Experts are encouraging in their assessments about the prospects for synergizing the methodologies. As noted by Charles L. Cell and Boris Arratia of the US Army Joint Munitions Command, "VE, Lean, and Six Sigma can work effectively, independent of the other methods, but they work better together, particularly in a process where a team can take advantage of respective strengths and avoid respective weaknesses."[†]

### 1. *Analysis of LSS*

The literature identified two primary areas for improvement of LSS: scope and the creative toolkit. As Kirkor Bozdogan of MIT notes, Six Sigma

[*] Referred to by some as Value Methodology (VM).

[†] Charles L. Cell and Boris Arratia, *Creating Value with Lean Thinking and Value Engineering* (Rock Island, IL: US Army Joint Munitions Command, 2003), 8.

lacks a "wide array of conceptually grounded and differentiated tools. Six Sigma discussions of change initiatives are designed to be quite localized and process specific."* Six Sigma's ability to identify and eliminate variation is useful for small-scale projects; however, it typically limits its examination of the life cycle in accounting for value. As a result, Six Sigma–based solutions are often narrow and project specific. Lean also has limitations in the scope of its solutions. While Lean is useful in identifying problems in existing products and processes, it is less applicable in developing new designs or solutions. Cell and Arratia note that "Lean principles and practices offer no direct method of addressing product design."† They also point out that Lean's success is often limited to high-value and high-cost projects that have sufficient management attention and support, including resources, to implement Lean recommendations.‡ However, that assessment is not to suggest that Lean and Six Sigma are lacking in unique strengths.

One of Six Sigma's greatest strengths is its use of data. Gordon Johnson of the International Truck and Engine Corporation argues that "The Six Sigma discipline allows an organization to make use of data through statistical analysis. It qualifies and quantifies the effectiveness of an operation and gives a means to continually improve the operation."§ These strengths highlight ways in which Lean and Six Sigma could synergize with VE. Six Sigma's collection techniques and use of data would contribute to Lean and VE studies. Henry Ball concludes that "As Six Sigma processes are data driven, the information derived is excellent input for a VM study."¶ Johnson also notes that Six Sigma's data and statistical methodology would be "useful in identifying problem areas as well as providing a way to quantify the functional impact of VM workshop proposals."**

As previously noted, Lean's broad approach in identifying waste is one of its greatest strengths and has the potential to contribute significantly to

* Kirkor Bozdogan, *Lean Aerospace Initiative: A Comparative Review of Lean Thinking, Six Sigma, and Related Enterprise Change Models*, Center for Technology, Policy, and Industrial Development (Cambridge, MA: MIT, December 3, 2003), 9.

† Cell and Arratia, *Creating Value with Lean Thinking*, 3.

‡ Cell and Arratia, *Creating Value with Lean Thinking*, 3.

§ Gordon S. Johnson, "Conflicting or Complementing? A Comprehensive Comparison of Six Sigma and Value Methodologies," in *SAVE International 43rd Annual Conference Proceedings* (Scottsdale, AZ, June 8–11, 2003), 5–6.

¶ Henry A. Ball, "Value Methodology—The Link for Modern Management Improvement Tools," in *SAVE International 43rd Annual Conference Proceedings* (Scottsdale, AZ, June 8–11, 2003), 7.

** Johnson, "Conflicting or Complementing?" 7.

a VE study. Bozdogan argues that "Lean thinking provides an overarching intellectual architecture for the various systemic change initiatives, wherein they augment each other in significant ways and represent mutually complementary approaches."[*] One of Lean's greatest contributions is the value stream map. The map includes detailed information about the process and is an effective tool for identifying areas of waste that have the greatest potential to improve effectiveness and efficiency and thereby create value.[†] Lean is particularly well aligned with VE because of the customer and value focus of both methodologies. As Cell notes, "Creating value is at the core of Lean. Creating value is at the core of Value Engineering. Lean and VE use different approaches to accomplish the same objective. Assuming we accept the idea that no one approach is superior … there may be concepts, approaches, and tools in each approach that could help the other."[‡]

John Sloggy comes to a similar conclusion from a product-development perspective. He states that "the best approach is to utilize the appropriate technique at the correct point in the product/project development cycle, as opposed to force fitting a specific process across all phases of the cycle."[§] He further describes places where LSS has limitations. "Lean Manufacturing, Six Sigma and TOC [Theory of Constraints] address the labor and variable overhead segments of the cost structure but have little or no impact on material cost, the largest segment of the pie. For the most part, design features of the product drive material and process costs, and the Lean/Six Sigma/TOC methodology offers little in the way of a tool kit for paring these costs. Because design features drive material and process costs, a comprehensive improvement effort must attack the material cost embedded in the product design. Supply chain development programs will reduce price (and material cost) to a degree, but these efforts will always be limited by the underlying characteristics of the product design."[¶]

---

[*] Kirkor Bozdogan, "Lean Aerospace Initiative: A Comparative Review of Lean Thinking, Six Sigma and Related Enterprise Change Models," Center for Technology, Policy and Industrial Development, (Cambridge, MA: MIT, December 3, 2003), 2.

[†] Cell and Arratia, *Creating Value with Lean Thinking*, 10.

[‡] Cell and Arratia, *Creating Value with Lean Thinking*, 6.

[§] John E. Sloggy, "The Value Methodology: A Critical Short-Term Innovation Strategy That Drives Long-Term Performance," in *SAVE International 48th Annual Conference Proceedings* (Reno, NV, June 9–12, 2008), 4.

[¶] Sloggy, "The Value Methodology," 4.

## 2. Potential VE Contributions

In a 2003 presentation to SAVE, Dr. Michael J. Cook identified six ways in which VE could contribute to Six Sigma: generate project ideas, develop business strategy, define problem/defect, identify root causes, generate improvements, and generate design concepts.* This study, which examines both Six Sigma and Lean, focuses on what makes VE unique as compared with LSS and what VE tools would be particularly useful in conjunction with an LSS project.

VE is rich in opportunities to broaden the scope of an improvement project. By examining the value of every function, VE captures a broad picture of a process while also offering solutions that will not detract from the customer-identified areas of highest value. In the Orientation phase, Cell notes that "VE's value approach and tools help teams focus on the high payoff areas first and will generate larger savings sooner than you might otherwise get in Lean."† Similarly, in the Creative, Evaluation, and Development phases, the methodology "offers analysts an effective analytic method for developing design changes to reduce cost and increase value."‡

Based on the literature, the Function Analysis System Technique (FAST) diagram is one of the strongest elements in the VE toolkit and has the greatest potential to contribute to LSS. FAST is the primary tool for gaining a broad perspective and identifying areas of improvement within an LSS project. The FAST diagram asks *why* and *how* questions that otherwise would be explored only on a more limited basis in an LSS project, as noted by Theresa Lehman and Paul Reiser of The Boldt Company.§ FAST is useful in almost all phases of the DMAIC and DFSS processes, particularly in the Define and Improve phases. FAST can provide a comprehensive view of an organization and break down processes based on functions and areas of the highest value. Focusing on functions ensures that high-value areas will be remedied for efficiency and effectiveness and wasting resources will be avoided. The literature also highlights the potential to apply FAST in the Design phase. Ball presents a theoretical application of FAST within an LSS project: Lean principles

* Michael J. Cook, "How to Get Six Sigma Companies to Use VM and Function Analysis," in *SAVE International 43rd Annual Conference Proceedings* (Scottsdale, AZ, June 8–11, 2003), 4.

† Cell and Arratia, *Creating Value with Lean Thinking*, 5.

‡ Cell and Arratia, *Creating Value with Lean Thinking*, 3.

§ Theresa Lehman and Paul Reiser, "Maximizing Value and Minimizing Waste: Value Engineering and Lean Construction," in *SAVE International 44th Annual Conference Proceedings* (Montreal, Quebec, July 12–15, 2004), 2.

identify wasteful activities in the production of a very complex part, FAST identifies the most valuable aspects of the part and its production, and FAST is used as a creative tool to make the complex part more producible.*

In discussing how FAST, coupled with creative techniques and supporting exercises, drives innovation, Sloggy argues that "Innovation is what separates high-performing organizations from the rest of the pack. Value Management provides the vehicle to accelerate past the competition and reestablish dominance in business. It is the right tool for the times, and, utilized in conjunction with the Six-Sigma/Lean/TOC quality focus, it provides a viable solution to today's intensive competitive challenges. From a public sector standpoint, the same benefits of a creative approach to problem solving provide unique solutions that are cost effective in these times of dwindling resources and conflicting priorities."†

## B. Observations and Analysis on Synergies‡

One observation, based on the comparison in Chapter 4 of the methodological approaches and the examples of synergies discussed in the literature, is that VE techniques are sometimes better equipped to lead to improvements or solutions complementary to those identified through a DMAIC/DFSS approach. These synergistic opportunities derive from the different perspective that VE takes in its Function Analysis and Creative phases as compared to the Analyze and Improve phases in DMAIC. Table 5.1 illustrates the differences in perspectives by showing the goal, focus, scope, change process, and business model for VE and the three components of LSS—Lean, Six Sigma, and TOC.§

The differences are small in the scope, change process, and business model rows of the table. However, the distinctions revealed in the relative goals and focus can be attributed to VE's more explicit consideration of cost and more active challenging of requirements that cost more than they are worth.

---

* Henry A. Ball, "Value Methodology," 4.

† John E. Sloggy, "The Value Methodology," 6.

‡ Some of the ideas in this section were derived from suggestions made by members of the Target Costing Best Practice Special Interest Group of the Consortium for Advanced Management International (CAM-I) in response to the discussion briefing: Jay Mandelbaum and Heather Williams, "Synergy in Enterprise Change Models: Opportunities for Collaboration Between Value Engineering and Lean/Six Sigma" (March 8, 2010).

§ Table 5.1 is modified from a brief developed by the Lean Advancement Initiative at MIT.

**Table 5.1 Comparison of VE and LSS Philosophical Approaches to Change**

| | *VE* | *Lean* | *Six Sigma* | *TOC* |
|---|---|---|---|---|
| Goal | Lower life-cycle cost and improve return on investment (ROI) | Eliminate waste | Reduce business risk* | Eliminate bottlenecks |
| Focus | Function analysis and function worth | All enterprise processes and people | All sources of product/ variation | Throughput |
| Scope | Enterprise | Enterprise value stream | Enterprise | Enterprise |
| Change process | Incremental product/process improvement | Evolutionary and systematic | Process specific, continuous | Continuous |
| Business model | Increase value to the stakeholder | Deliver value to all stakeholders | Minimize waste and increase customer satisfaction | Increase financial performance of core enterprise |

* The original MIT entry for this cell was "reduce variation in processes." The latest thinking is that Six Sigma is primarily concerned with process improvement, but not all process improvement is focused on variation reduction.

In Table 5.1, VE's goal explicitly considers cost; consequently, VE does not suboptimize from a financial perspective. VE only develops alternatives that provide the necessary functions, and the cost of every alternative is estimated using a cost model. The total cost estimate (life-cycle cost) is then considered (along with other things such as schedule and feasibility) in a decision-making process. If something else is being optimized (e.g., reduction in variation or waste or the elimination of bottlenecks), the effect on total cost is unclear. For example, rework costs may decrease while production costs may increase. Both VE and LSS are necessary to reduce costs—VE focuses on what is done (i.e., the function) and LSS improves how it is done (i.e., with minimal waste).

VE is not concerned simply with cost reduction. Its primary focus is on spending only what is necessary to meet the requirements, thereby yielding improved value and return on investment (ROI). Since understanding requirements is inextricably linked to delivering functions, any alternative that provides the necessary functions in a FAST diagram will automatically satisfy the customer's requirements. An alternative that provides something

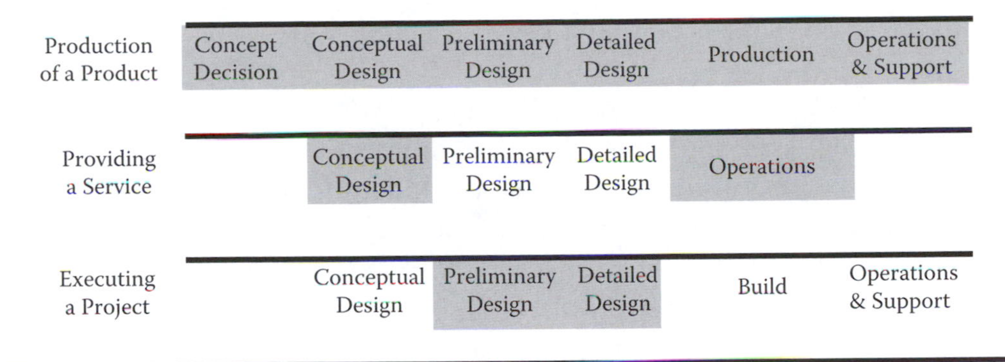


**Figure 5.1 High-leverage opportunities for VE throughout the life cycle.**

beyond the necessary functions is probably providing something that is not highly valued by the customer—a cost–value mismatch. Therefore, function analysis challenges requirements by ensuring that areas of major expenditure receive attention through critical and innovative thinking about alternatives. This does not always happen by reducing variation or waste or increasing throughput. In addition, challenging the requirements discourages changing the requirements at a later time.

Figure 5.1 illustrates the principal areas where these differences in perspective enable VE to augment LSS efforts from a product, service, or project life-cycle perspective.

## *1. Synergies in Producing a Product*

VE concepts, methods, and applications provide benefits across all phases of a product's life cycle and have been successfully applied in the Department of Defense (DOD), where a concept decision determines an overarching approach to meet a capability need. The approach can include any combination of doctrine, organization, training, materiel, leadership and education, personnel, or facilities (DOTMLPF).[*] If a materiel solution is pursued, an Analysis of Alternatives (AoA) assesses the potential materiel solutions to satisfy the capability need.[†] By considering function and cost, a VE approach can provide important insights, and function analysis determines what must be done. Brainstorming in the Creative phase considers all DOTMLPF

[*] See Chairman of the Joint Chiefs of Staff Instruction (CJCSI) 3170.01G, *Joint Capabilities Integration and Development System* (March 1, 2009).

[†] US Department of Defense, DOD Instruction (DODI) 5000.02, *Operation of the Defense Acquisition System* (December 8, 2008).

options to accomplish those functions. LSS is almost never used this early in a product life cycle.

While DFSS is a proactive and anticipatory approach that helps evaluate and optimize conceptual, preliminary, and detailed designs, it is not an automatic process and does not replace skilled designers. Developing an effective design that does everything a user wants from a performance perspective and from the perspective of design considerations (e.g., supportability, maintainability, information assurance, availability, reliability, producibility may be applicable) while not costing too much or weighing too much will almost always benefit from the group perspectives and discussions of the Function Analysis and Creative phases of the VE job plan. VE links the customer requirements to the design to manage cost. Companies worldwide integrate VE concepts into their design processes to establish target costs and ensure that unnecessary functions and requirements are eliminated.

The following simple FAST example illustrates how VE can be used in design. Figure 5.2 is the actual FAST diagram developed in a VE workshop for the Army's Improved Chemical Agent Monitor (ICAM). Figure 5.3 is a picture of the ICAM. The ICAM dust cap can be seen in the upper left corner of Figure 5.3. It is attached to the ICAM unit by a plastic strap.

The boxes numbered 4–6 and 10–14 in Figure 5.2 described how the ICAM is operated. When the dust cap is removed to use the ICAM, it should remain attached by the plastic strap. The installation of the dust cap is shown in boxes 3 and 7–9 in Figure 5.2. The problem being experienced

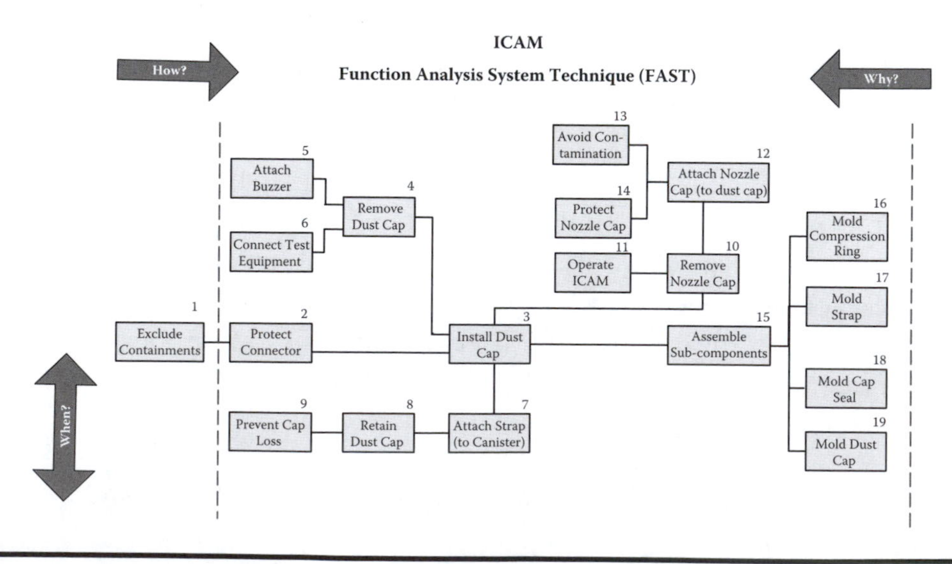


**Figure 5.2 FAST diagram for the Army's ICAM.**

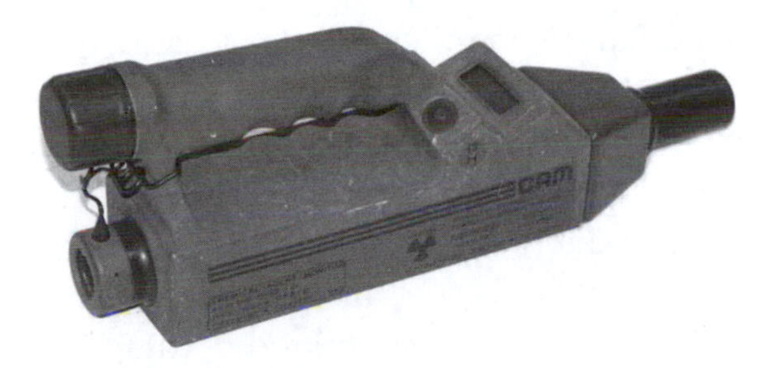


**Figure 5.3 Improved Chemical Agent Monitor (ICAM).**

in the field was that the strap was prone to breaking, and when that happened, the dust cap was usually lost. Replacing the dust cap is costly since the strap, dust cap, compression ring, and seal are a single plastic-injected part (see boxes 15–19 in Figure 5.2). To correct this problem, 36 ideas were developed during the VE brainstorming session. Two of them were developed further: (1) make the strap a separate part and (2) use coated braided wire for the strap and attached it to the dust cap assembly with a rivet.

VE has little to add to LSS in improving manufacturing processes on the production floor; however, VE can contribute to the production phase of a product's life cycle in other ways. Production costs can often be reduced by introducing new technologies, new processes, new materials, and/or new designs. The VE methodology can more readily identify creative problem-solving approaches than LSS, especially in government contracts.*

A real-world example of this is the Phalanx Close-In Weapon System (CIWS), a fast-reaction, rapid-fire 20-mm gun system that provides Navy ships a terminal defense against antiship missiles and fixed-wing aircraft that have penetrated other fleet defenses. Phalanx uses advanced radar and computer technology to locate, identify, and direct a stream of armor-piercing projectiles to the target.

A contract was awarded to retrofit Phalanx with a manual controller to direct fire against targets of opportunity. Using the Function Analysis phase of the VE methodology, the contractor identified an opportunity to replace a

* While the government encourages both VE and LSS, some special implementation requirements are associated with VE. In 1993, OMB Circular A-131 mandated the use of VE by all federal agencies and also required annual reporting on VE savings. This mandate was reinforced by the National Defense Authorization Act of Fiscal Year 1996 (Public Law 104-106, Section 4306) and updated by Public Law 11-350 in January 2011. The Value Engineering section states, "Each executive agency shall establish and maintain cost-effective procedures and processes for analyzing the functions of a program, project, system, product, item of equipment, building, facility, service, or supply of the agency. The analysis shall be (1) performed by qualified agency or contractor personnel; and (2) directed at improving performance, reliability, quality, safety, and life cycle costs."

military standard fixed-hand controller (similar to a joy stick) with a derivative of a commercial unit not built to military standards.* The contractor, on its own initiative, worked with the commercial source to produce a modified unit and tested the unit against the requirements for the military standard version. Based on the test results, the contractor had confidence that the commercial derivative would meet all of the technical requirements at a lower cost. Therefore, the contractor submitted a Value Engineering Change Proposal (VECP)† to replace the standard military controller with ruggedized commercial derivatives. The military standard controller cost $7,600. The commercial derivative cost only $2,100. Since each gun required three controllers, the net savings would be $16,500 per system. The US Navy and the contractor shared approximately $2 million in savings. Eventually, the Navy can save more than $9 million if the idea is applied to all ships. In addition, the VECP provided for earlier implementation of the improved system.

This example illustrates a second point. The VE methodology can develop additional alternatives, and the contractual use of VECPs with the government can create incentives for the contractor to develop new ideas, thereby creating further opportunities for synergy.

VE can also attack variation in production differently than LSS. For example, when Alan Mulally became CEO of Ford Motor Company in 2006, he targeted unnecessary and costly product variations that contributed no value to the customer. The following anecdote was not actually a VE application, but it illustrates the kind of solution that might be derived from VE. According to the *Wall Street Journal*, Mulally "laid out 12 different metal rods that Ford uses to hold up a vehicle's hood. He wanted to demonstrate to managers that this kind of variation is costly but doesn't matter to customers."‡

In the operations and support phase of the product life cycle, VE and especially VECPs provide additional opportunities to enhance LSS-developed

---

* This example illustrates the power of function analysis in indentifying alternative (less costly) ways to perform required functions.

† VE has two implementation mechanisms. A Value Engineering Proposal (VEP) is a specific proposal developed internally for total value improvement from the use of VE techniques. Since VEPs are developed and implemented internally, all resulting savings accrue to the implementing organization. A VECP is a proposal submitted to the government by the contractor in accordance with the VE clause in the contract. A VECP proposes a change that, if accepted and implemented, provides an eventual, overall cost savings to the government. The contractor receives a substantial share in the savings accrued as a result of implementation. It, therefore, provides a vehicle through which acquisition and operating costs can be reduced while the contractor's rate of return is increased.

‡ Monica Langley, "Inside Mulally's 'War Room': A Radical Overhaul of Ford," *Wall Street Journal*, December 22, 2006.

options. Within the diminishing manufacturing sources and material shortages (DMSMS) situation, VE concepts can identify a large number of resolution options, evaluate their potential for solving the problem, develop recommendations, and provide incentives for the investments needed for successful implementation.* Using the VE methodology provides greater opportunity for developing and implementing innovative solutions to DMSMS problems.

For example, a defense missile contractor had a sole-source subcontractor for a costly warhead. The subcontractor was having problems meeting *insensitive munitions capability* requirements for the warhead not to explode in a fire or if dropped. With the cooperation of the government, the contractor submitted a VECP to develop an alternative, less-expensive source for the warhead by using reverse engineering. Since a different manufacturer would be used, the performance of the warhead's insensitive munitions capability could also be improved because this manufacturer would use a different process for making the explosive portion of the warhead. Approximately $12 million is currently being invested to develop the new source. Although savings of $15,000 per warhead is expected, the development of the second source makes this VE change and development of a second source even more valuable. Without the competition from another source, the price of the warhead probably would have continued to escalate as it had in the past since the single source had no incentive to control costs.

## 2. *Synergies in Providing a Service*

VE application to the design or redesign of a service (and by analogy a process) is similar to the product situation. The following hypothetical example† assumes a 3-year contract (a base year plus 2 option years) for the professional services of a physician to give full physicals to 3,600 military personnel each year (i.e., 10,800 physicals [3,600 × 3]) for $100 each, for a cost of $1,080,000. The associated contract requirements reflected in

* Jay Mandelbaum, Royce R. Kneece, and Danny L. Reed, *A Partnership between Value Engineering and the Diminishing Manufacturing Sources and Material Shortages Community to Reduce Ownership Costs*, IDA Document D-3598 (Alexandria, VA: Institute for Defense Analyses, September 2008).

† Jay Mandelbaum, Ina R. Merson, Danny L. Reed, James R. Vickers, and Lance M. Roark, *Value Engineering and Service Contracts,* IDA Document D-3733 (Alexandria, VA: Institute for Defense Analyses, June 2009).

**Table 5.2 Medical Service Contract Example before VECP Changes**

| *CLIN* | *Description* | *Quantity* | *Unit* | *Unit Price* | *Total Price* |
|---|---|---|---|---|---|
| 0001 | Provide a complete annual physical to military personnel | 10,800 | EA | $100 | $1,080,000 |

Table 5.2 depict what a contract line item (CLIN) may look like.* Figure 5.4 is a functional analysis representation of the situation.

Using VE to challenge the requirements creates opportunities to improve upon LSS solutions. In the preceding example, most of these military personnel are young and in excellent physical condition; therefore, the

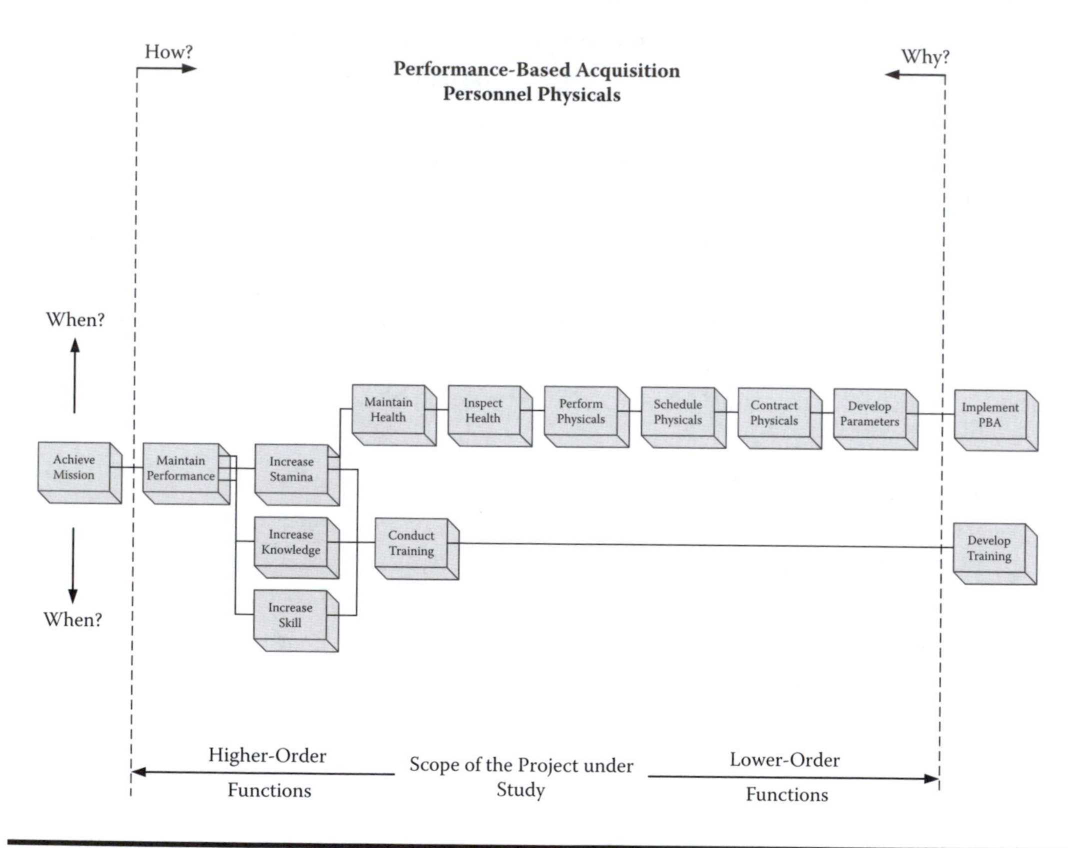


**Figure 5.4 VE function analysis on a services contract to provide physicals before a VECP.**

* This example depicts only one element of a larger contract. Obviously, some people would need more extensive medical care as a function of their physical condition. Such care would be provided in a separate CLIN. Also, depending on a person's occupation, additional assessments may be required. This example focuses only on that element of the population required to have a physical as their annual health assessment.

contractor could propose a VECP for a modified physical plan. Under the plan, anyone under 25 years of age would get a complete physical every 3 years, anyone 26–35, every 2 years, and anyone over 36, every year.* Those personnel not given a complete physical would receive a modified physical that could be performed at a lesser cost of $50. The VECP results shown in Table 5.3 assume the military population is divided equally among the three age bands. Figure 5.5 shows the corresponding function analysis.

Function analysis challenges requirements by questioning the existing system, encouraging critical thinking, and developing innovative solutions. It ensures that areas of major expenditure receive attention in the early stages of a service contract. The government receives substantial benefits—costs in this example are reduced by more than 10%.† Without VE, however, the contractor does not have an incentive to propose such a requirements change. Since a contractor needs some incentive to perform less work, a

**Table 5.3 Medical Service Contract Example after VECP Changes**

| *CLIN* | *Description* | *Quantity* | *Unit* | *Unit Price* | *Total Price ($)* |
|---|---|---|---|---|---|
| 0001 | Provide a complete annual physical to military personnel | 6,000 | EA | $100 | 600,000 |
| 0002 | Provide a modified physical to military personnel | 4,800 | EA | $50 | 240,000 |
| | **Subtotal** | **10,800** | | | **840,000** |
| | VECP Savings ($1,080,000–$840,000) | 240,000 | | | |
| | Contractor's Share of Savings Using a 50/50 share ($240,000 × .5) | 120,000 | | | |
| 0003 | New CLIN for VECP savings | 10,800 | EA | $25.00 | 120,000 |
| | **New Contract Total** | | | | **960,000** |

* This example is not intended to imply that the military would ask for more service than it needs. Instead, it illustrates how risk/requirements trades can be made.

† A secondary issue is that indirect rates may have to be increased if the reduction in the number of billed hours is significant.

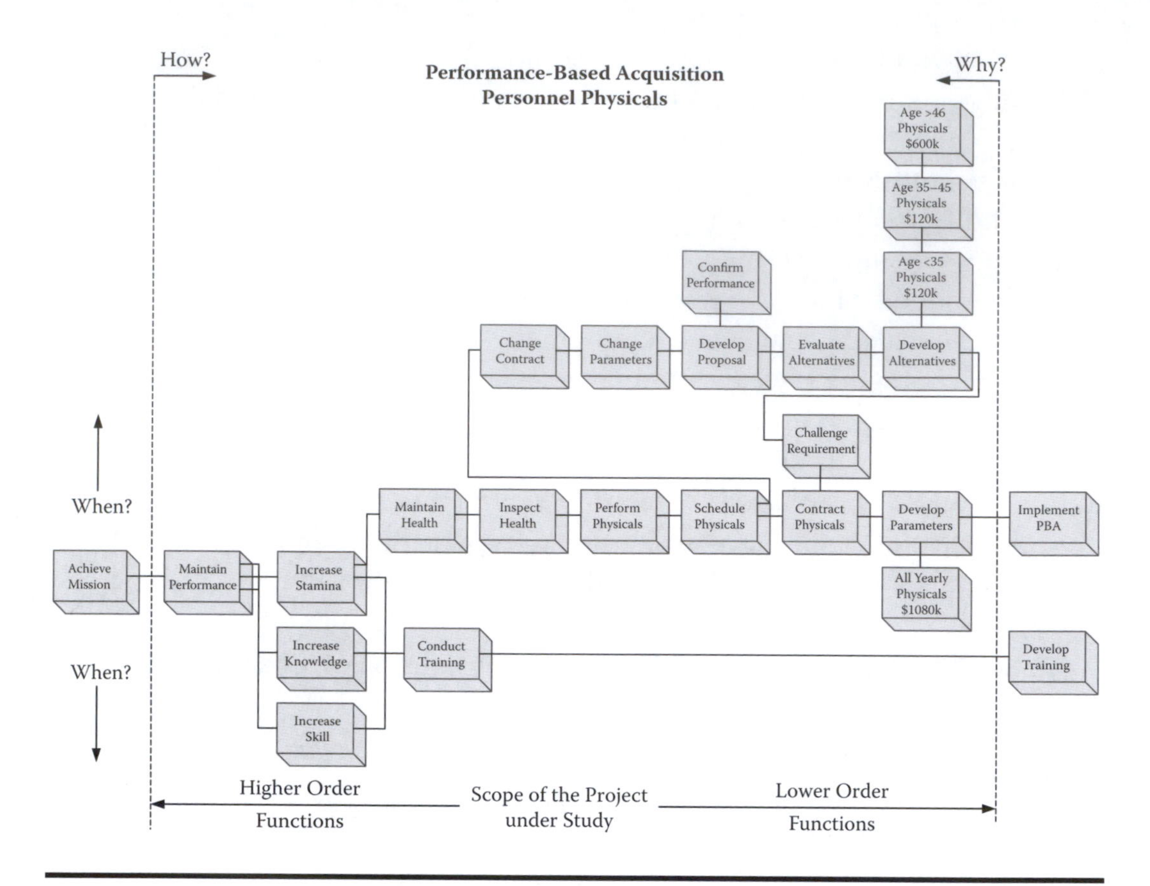


**Figure 5.5 VE function analysis on a services contract to provide physicals after a VECP.**

better mechanism of compensation is needed for the contractor to propose a VECP in a service environment. When less work is performed, revenue is down, so a balance or trade-off to increase profit must be found to make the change a worthwhile proposition for the contractor.* In this example, if a 10% profit is assumed, the $120,000 share of the savings appears to more than compensate for lost revenue. In the next contract, implementation of this idea would provide 100% of the savings to the government. The government also allows the contractor to receive a 10% royalty for use of his idea for the next three years.

* Collateral savings are also associated with the VECP depicted in Table 5.3. Since the modified physicals take less time, people would not be away from work as long and, therefore, would be able to perform additional duties. Since this benefit is relatively small and difficult to quantify, such collateral savings are normally not claimed.

## *3. Synergies in Executing a Construction Project*

VE has the greatest potential to augment LSS when the approach and costs are known (e.g., after design definition, approved feasibility study, and early remedial design are completed). As in all of the preceding discussions, VE identifies the essential functions and derives lower-cost alternative ways of accomplishing them. Both VEP and VECP approaches apply. The VE Workshop is an opportunity to bring the design team and client together to review the proposed design solutions, the cost estimate, and the proposed implementation schedule and approach, with the goal of achieving the best value for the money. The definition of what is good value on any particular project will change from client to client and project to project.

For example, VE has been used to make the following changes in DOD construction projects:

- Construct slab-on-grade in lieu of a more expensive structural slab where feasible. The former option would save money because it requires less effort and is only necessary to create a hard grade with a bulldozer. If a structural slab were needed, digging below the frost grade, leveling, and recompacting would be required to install footings.
- Combine buildings or phases of construction to reduce site preparation costs and allow utilities and heating, ventilation, and air conditioning (HVAC) systems to be shared.
- Use electrical generators for power peak-sharing, which would allow building occupants to stay below the peak rates. By staying below the peak rate, savings are accrued throughout the year.
- Control the HVAC system with a direct digital instead of a single-loop system. Since HVAC is on a timer and a digital timer is more accurate and can have better temperature control, this approach would eliminate overcooling, and the HVAC system would be used only when people are in the building.
- Use a waste heat recovery system. Money is saved when recycled heat generated by one process is used in another.
- Install alternative insulation behind precast concrete bonds. A concrete bond would be made in a factory and simply put in place. This could reduce labor cost and is also a faster way to apply insulation.

VECPs are more applicable when the construction project has separate design and build contracts. Contractors are provided monetary incentives to propose solutions that offer enhanced value to the government and share in the financial benefits realized. Clearly, the government must consider contractor-generated proposals carefully from a life-cycle and a liability perspective. The architect and engineer teams must be part of the decision-making process to ensure that the proposed change does not have any negative impact on the overall design and building function. VECP evaluation is treated similarly to any change order during construction, with issues such as schedule and productivity impacts being considered along with the perceived cost savings generated. As a result, the functionality of the project is improved, costs are thoroughly checked and reduced over the life cycle, and a second look at the design produced by the architect and engineers gives the assurance that all reasonable alternatives have been explored.

# Chapter 6

# Final Remarks

Both Lean Six Sigma (LSS) and Value Engineering (VE) have unique attributes and perspectives for process improvement. Since certain problems can be more readily, effectively, or thoroughly managed by using one or both of these perspectives, exploring the full range of solution options is crucial. In some circumstances, VE techniques are better equipped to lead to improvements or solutions that complement those identified through a DMAIC/DFSS (Define, Measure, Analyze, Improve, and Control/Design for Six Sigma) approach. These opportunities for synergy include the following:

- **Function analysis and the Function Analysis System Technique (FAST) diagram.** The disciplined use of function analysis is the principal feature that distinguishes the VE methodology from other improvement methods. Function analysis challenges requirements by questioning the existing system, encouraging critical thinking, and developing innovative solutions.
- **Cost focus.** VE only develops alternatives that provide the necessary functions. By examining only those functions that cost more than they are worth and identifying the total cost of each alternative, VE explicitly lowers cost and increases value.

VE does not take the place of LSS efforts, but it does present significant opportunities to enhance LSS-developed options. Therefore, it would be a good idea to augment LSS training to include the VE approach to function analysis, creativity, and associated elements of evaluation and development to identify candidate solutions as part of the Analyze and Improve phases of DMAIC.

To this end, Appendices C, D, and E provide suggested VE training material that has been used by the Office of the Secretary of Defense in its LSS Black Belt training. There are two versions of the training—one for products (and construction) and the other for a process application. Appendix C is the common front end for both versions. Appendix D is the remainder tailored for a product, and Appendix E is the remainder tailored for a process.

As far as DFSS is concerned, VE tools should be explicitly used in the process. They should be used in the Analyze phase of Define, Measure, Analyze, Design (and Optimize), and Verify (DMADV) to construct function views of the product or process to identify customer priorities and determine functional requirements. They should also be used in the Design phase of DMADV to generate alternative design concepts and to modify component/subsystem preliminary and detailed designs to introduce new elements to the evaluation and optimization processes.

Similarly, VE practitioners should incorporate several LSS features when they prepare for and conduct workshops. Examples include the use of the following:

- Customer communication tools such as Likert scales, surveys, interviews, and focus groups to set goals early in the process
- A more formalized data collection plan
- Supplies, Inputs, Process, Outputs, and Customers (SIPOC) to add insight during function analysis
- A control plan to ensure implementation proceeds as planned
- A formal corrective action plan to adapt to changes during implementation

# Appendix A: Cross-Reference Charts in Detail

Gray boxes represent the VE job plan, and white boxes represent the DMAIC methodology. The activities represented by diagonal stripes identify structural differences (in the Define, Measure, and Control phases), and the activities with dark gray shading represent a difference in the analytical approach (in the Analyze and Improve phases).

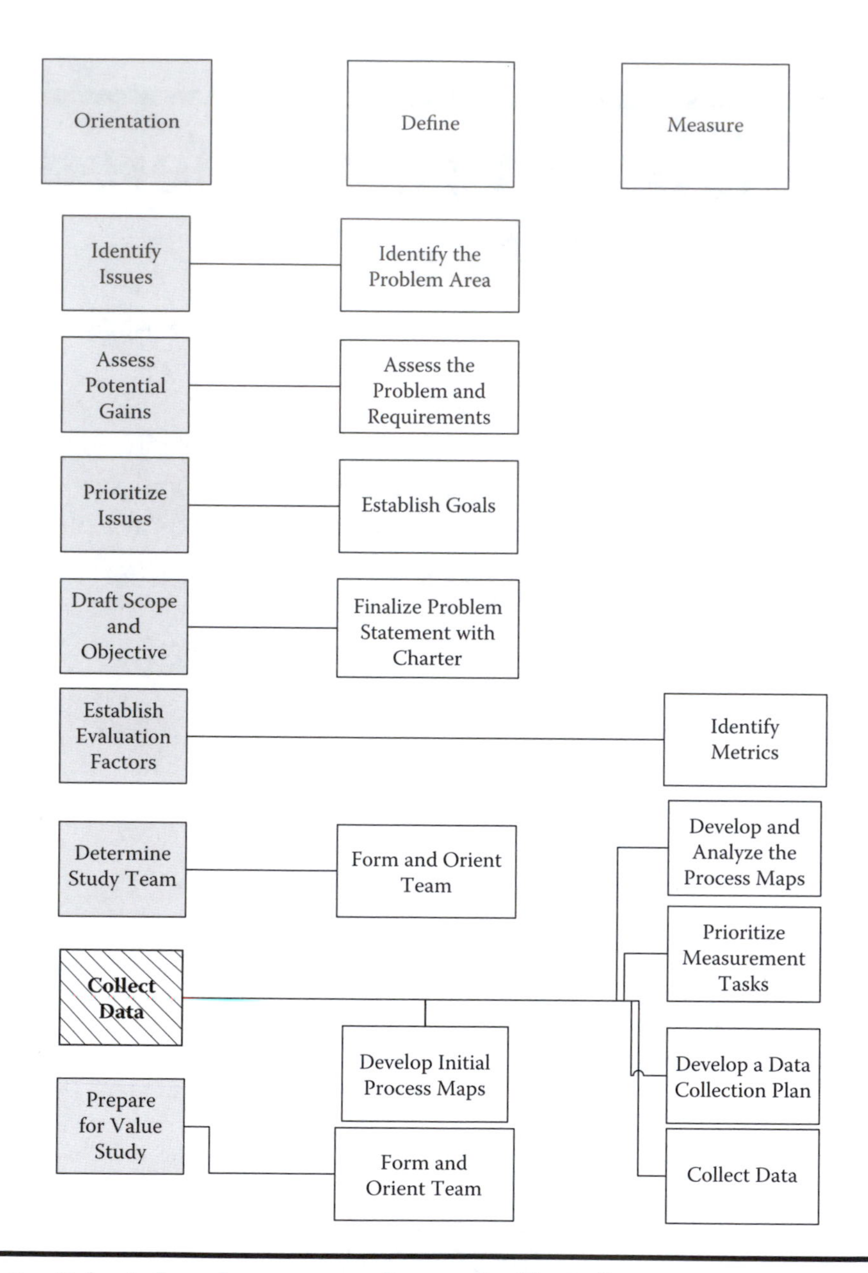


**Figure A.1 Orientation phase cross-referenced with Define, Measure, Analyze, Improve, and Control (DMAIC) methodology.**

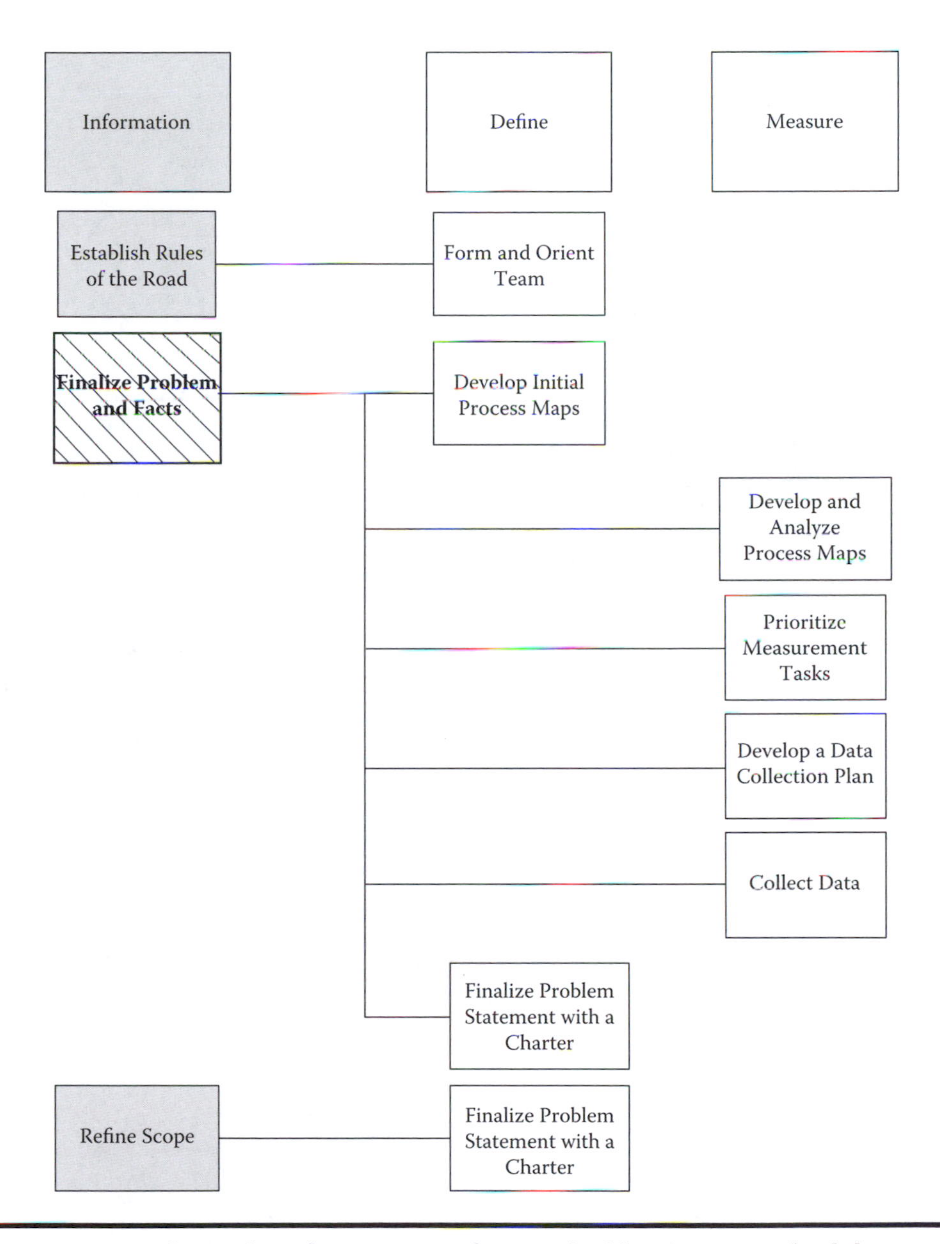


**Figure A.2 VE Information phase cross-referenced with DMAIC methodology.**

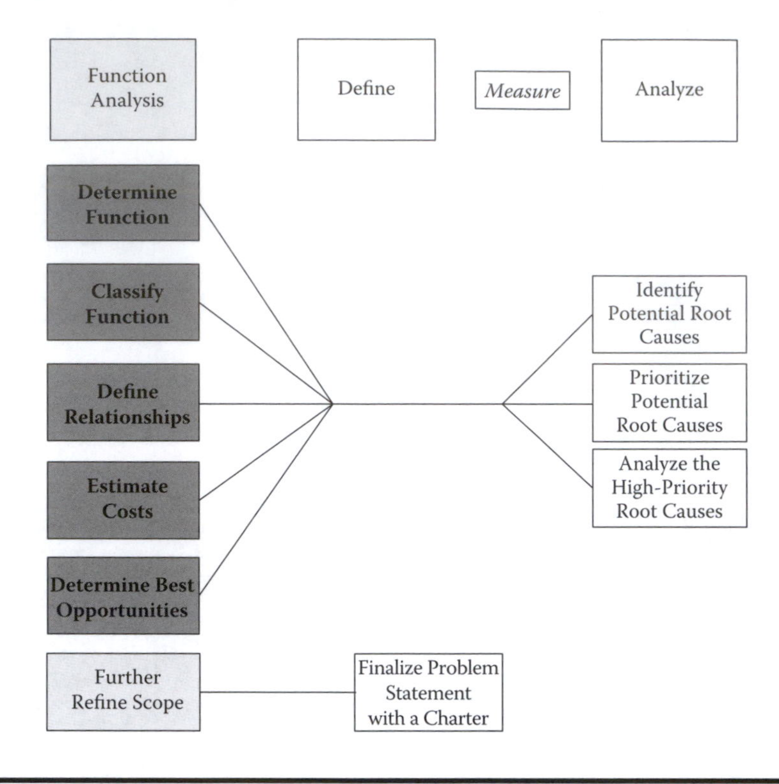


**Figure A.3 VE Function Analysis phase cross-referenced with DMAIC methodology.**

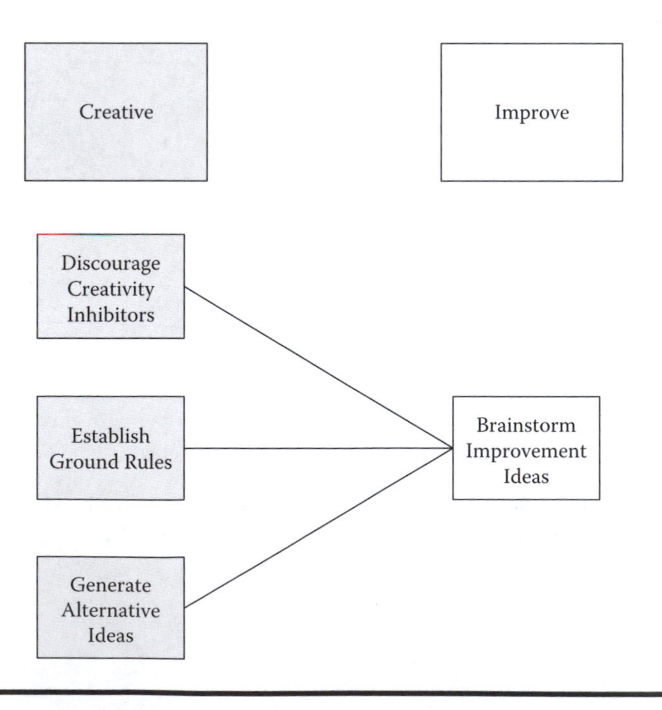


**Figure A.4 VE Creative phase cross-referenced with DMAIC methodology.**

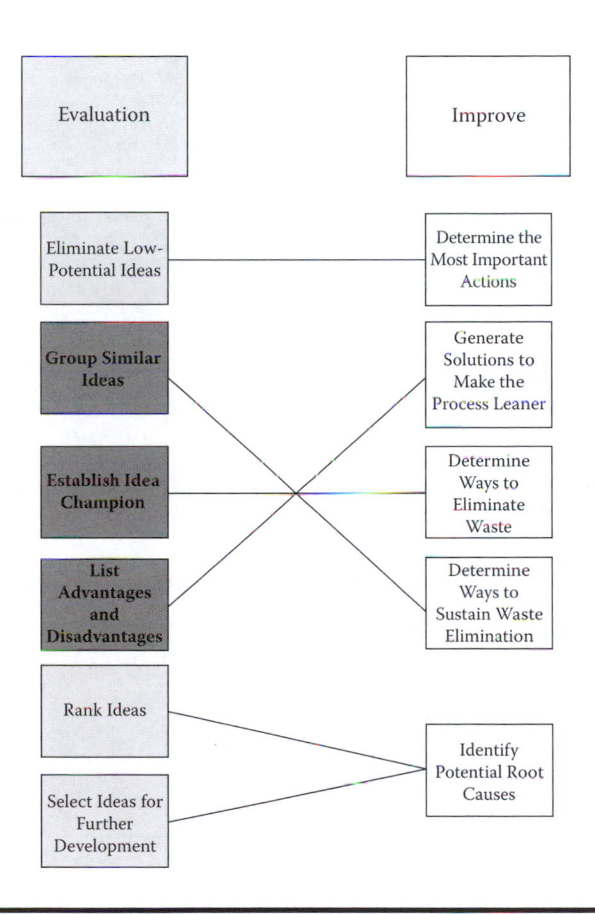


**Figure A.5 VE Evaluation phase cross-referenced with DMAIC methodology.**

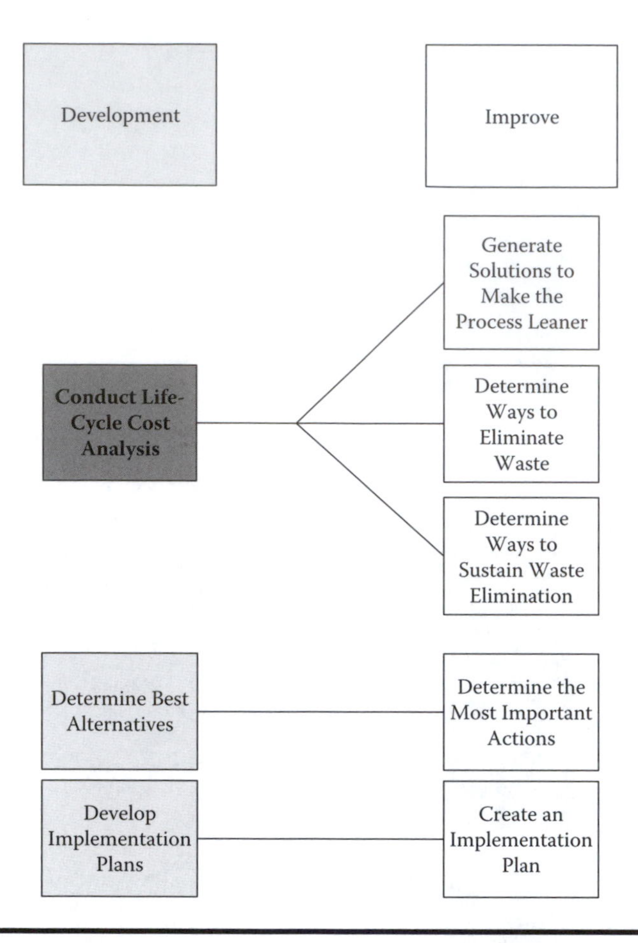


**Figure A.6 VE Development phase cross-referenced with DMAIC methodology.**

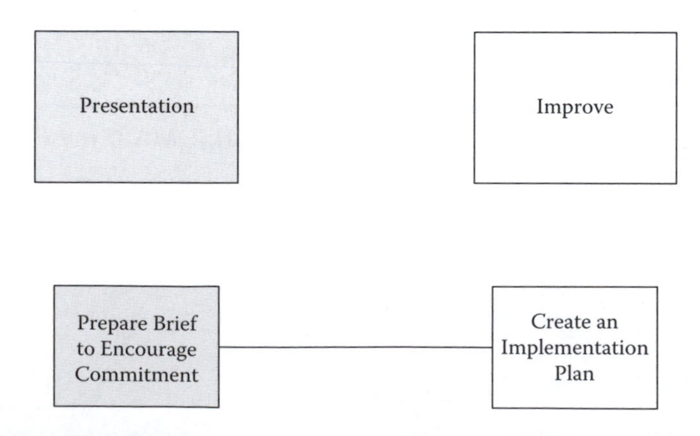


**Figure A.7 Presentation phase cross-referenced with DMAIC methodology.**

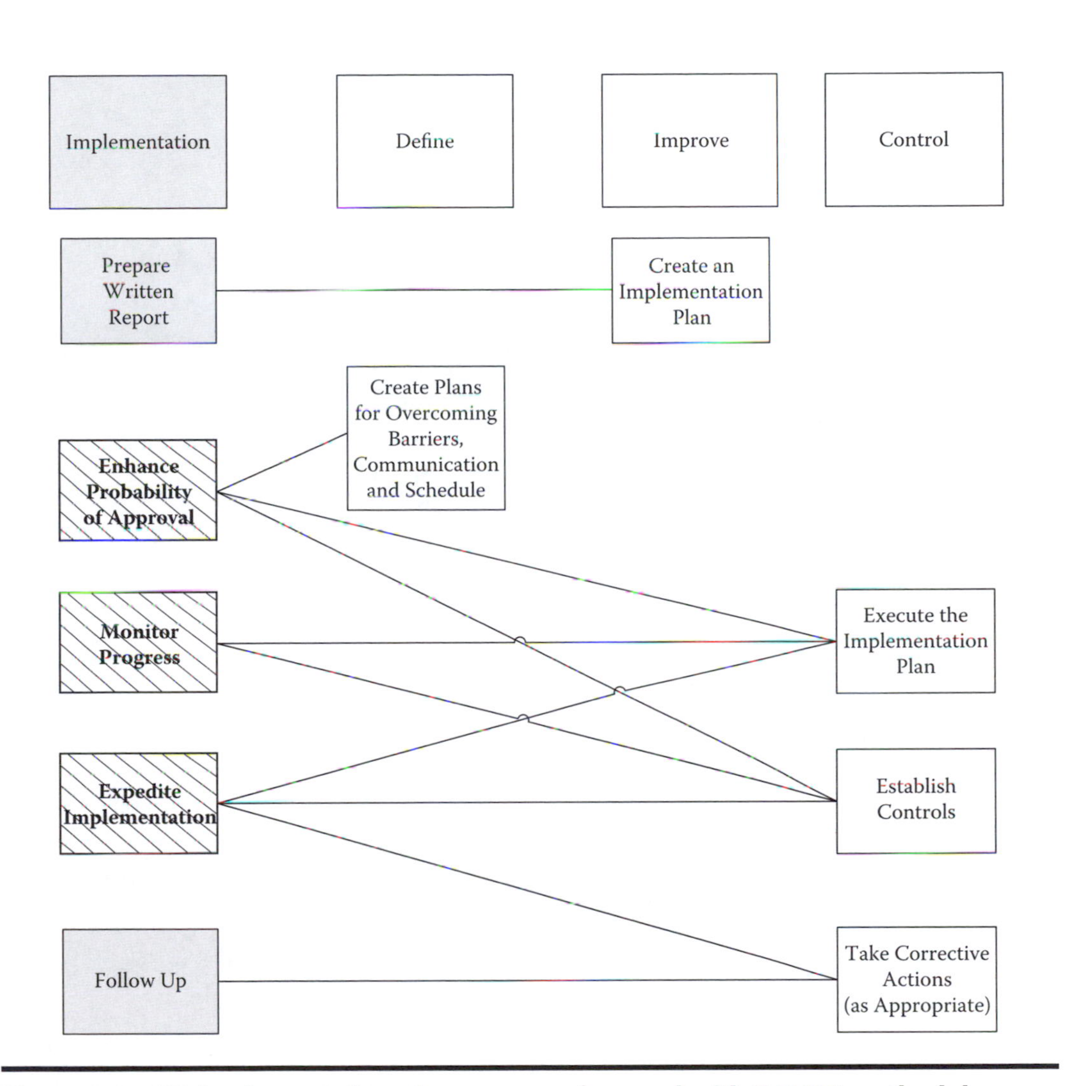


**Figure A.8 VE Implementation phase cross-referenced with DMAIC methodology.**

# Appendix B: Common LSS, DFSS, and VE Tools

**Table B-1 Common Lean Six Sigma (LSS) Tools**

| Define Phase |
|---|
| Interviews |
| Focus Groups |
| Surveys |
| Focused Project Definition Tree |
| Top-Down Flowchart |
| Spaghetti Flowchart |
| Detailed Flowchart |
| Deployment Flowchart |
| Time Value Chart |
| SIPOC |
| **Measure Phase** |
| Quality Function Deployment (QFD) |
| Gage Repeatability and Reproducibility (GRR) |
| Short Method |
| Test Retest |
| Attribute GRR |
| P(Miss), P(False Alarm), Overall Effectiveness |

| Analyze Phase |
|---|
| Benchmarking |
| Solution Tree |
| Normality Plot |
| Histogram |
| Run Chart |
| Capability Analysis Normal |
| Six Sigma Process Report |
| Six Sigma Product Report |
| Solution Tree |
| Progressive Search |
| Multi-Vari Chart |
| One-Sample T |
| Two-Sample T |
| Paired T |
| One-Way Analysis of Variance |
| Homogeneity of Variance |
| General Linear Model |
| Power and Sample Size |

| **Analyze Phase** (cont.) |
|---|
| Dot Plot |
| Box Plot |
| Pareto |
| Mood's Median |
| Runs Test |
| Chi-Square Minitab |
| Chi-square Excel |
| Phi Statistic |
| Fisher's Exact |
| Binary Logistic Regression |
| Power and Sample Size |
| Scatter Plot |
| Correlation |
| Simple Linear Regression |
| Multiple Linear Regression |
| Polynomial Regression |
| Matrix Plot |
| Residual Plots |
| Binary Logistic Regression |
| Component Search |
| Paired Comparisons |

| **Improve Phase** |
|---|
| Tolerance Parallelogram |
| Crystal Ball Simulation |
| Design of Experiments |
| Full Factorial |
| Fractional Factorial |
| **Control Phase** |
| Failure Mode and Effects Analysis |
| Mistake Proofing |
| Control Charts |
| Process (Low Volume) |
| Pre control |
| Control Plan |
| **Lean** |
| Value Stream Map |
| Waste |
| Pull |
| Kanban |
| Takt Time |
| Standard Work Combination Sheet |
| Standard Work In Process |
| Material Presentation |
| Visual Management |

**Table B-2 Common Design for Six Sigma (DFSS) Tools**

| **Define** |
|---|
| Multi-Generation Product Plans |
| Multi-Generation Technology Plans |
| Kano model |
| QFD |
| Customer Surveys |
| Affinity Diagrams |
| **Measure** |
| Behavior Models |
| Context Diagrams |
| Structure Tree |
| Critical To Quality (CTQ) Flowdown |
| Target Costing |
| Benchmarking |
| Measurement Systems Analysis/GRR/ Calibration |
| **Analyze** |
| Brainstorming |
| Benchmarking |
| Pugh Concept Selection |
| Weibull Reliability Analysis |
| Systems Reliability Scorecards |
| Hypothesis Testing |
| DFSS Scorecards |
| Failure Mode Effects and Criticality Analysis |
| FMEA/FMEA Lite |
| Risk Assessment |

| **Design** |
|---|
| Design and Analysis of Computer Experiments |
| Design of Experiments |
| Metamodeling—Regression |
| Non-Parametric Metamodeling |
| Process Capability Analysis |
| Process Capability Databases |
| Data Visualization Tools—Data Mining |
| **Optimize** |
| Robust Design/Reliability |
| Optimization—Derivative Based |
| Optimization—Stochastic |
| Filtered Monte Carlo Optimization |
| Multi-Objective Optimization |
| Partial Derivatives |
| Fast Probability Integration |
| Point Estimate Method |
| Worst Case Tolerance Analysis |
| Root Sum Squares Tolerance Analysis |
| 3D Tolerance Analysis |
| Error Proofing |

**Table B-3 Common Value Engineering (VE) Tools**

| **Information Phase** |
|---|
| QFD |
| Voice of Customer |
| Strengths, Weaknesses, Opportunities, and Threats (SWOT) |
| Project Charter |
| Benchmarking |
| Design for Assembly |
| Pareto Analysis |
| Tear-Down Analysis |
| **Function Analysis Phase** |
| Random Function Identification |
| Functional Analysis System Technique (FAST) |
| Function Tree |
| Cost to Function Analysis (Function Matrix) |
| Failure Modes and Effects Analysis (FMEA) |
| Performance to Function Analysis |
| Relate Customer Attitudes to Functions |
| Value Index |
| **Creative Phase** |
| Creativity "Ground Rules" |
| Brainstorming |
| Synetics |
| Theory of Inventive Problem Solving (TRIZ; in Russian, *Teoriya Resheniya Izobretatelskikh Zadatch*) |
| Nominal Group Technique |
| Gordon Technique |
| **Evaluation Phase** |
| T-Charts |
| Pugh Analysis |

**Table B-3 Common Value Engineering (VE) Tools (Cont.)**

| |
|---|
| Value Metrics |
| Choosing by Advantages |
| Life-Cycle Costing |
| Kepner-Tregoe |

# Appendix C: General VE Material for LSS Black Belt Training

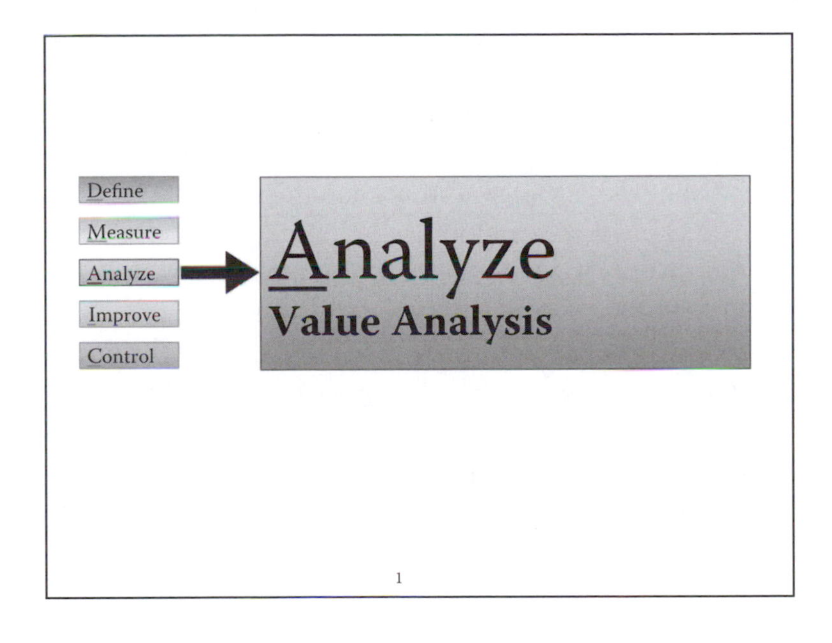

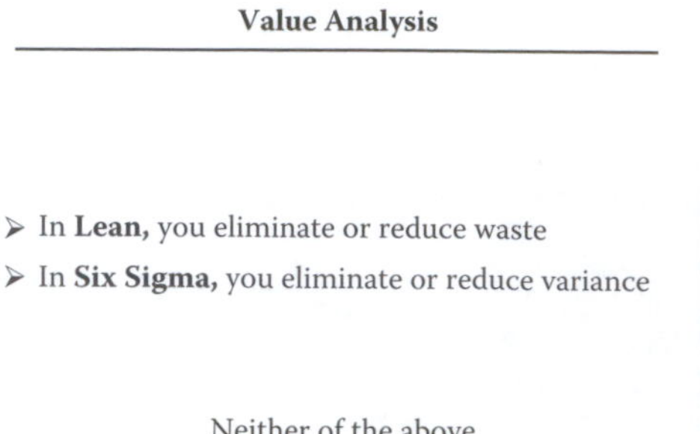


We will use the simple term *product* where product could be systems, components of systems, procedures, regulations, processes, construction or items of commerce.

Value Analysis gives you the opportunity to change a product by performing the product's functions in an alternative manner.

In doing just Lean and Six Sigma, there is a tendency to focus on the labor, materials, and processes being used to make a product and strive for improvement in a given situation. In Six Sigma you can reduce variance in one of two ways: improve the process capability or increase (loosen) the tolerance of the design to fit within the process capability. Although the product may change, the focus is on the process to build the product, not the product itself.

Value Analysis allows you to break out of that mold. Why improve something when changing course has the potential to make even greater improvement?

**Value Analysis**

- A planned effort
- to analyze **functions** of
- systems, designs, criteria, procedures
- to satisfy all needed user requirements
- at lowest total cost of ownership!

3

Just as you learned in week 1 of this course for Lean and Six Sigma, Value Analysis requires a planned effort.

However, its methodology is to analyze the function or functions your product is expected to achieve. Because everything has a function (or you would not be spending resources on it), everything is susceptible to value analysis.

The purpose of value analysis is to satisfy all NEEDED user requirements and do that at the lowest total cost of ownership. In other words, *life-cycle cost.*

**Instructor note:** Total cost of ownership means life-cycle cost.

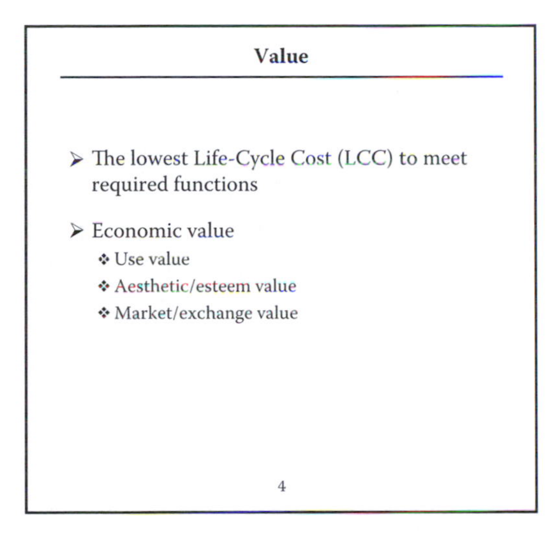


Value Analysis only deals with Economic Value.

We do not deal with Social Value, Political Value, Judicial Value, or Moral Value.

There are three types of Economic Value:

1. *Use value* relates to the work function of a product.
2. *Aesthetic* or *esteem value* relates to the sell function of a product.
3. *Market* or *exchange value* comes into play at the end of the product's life cycle.

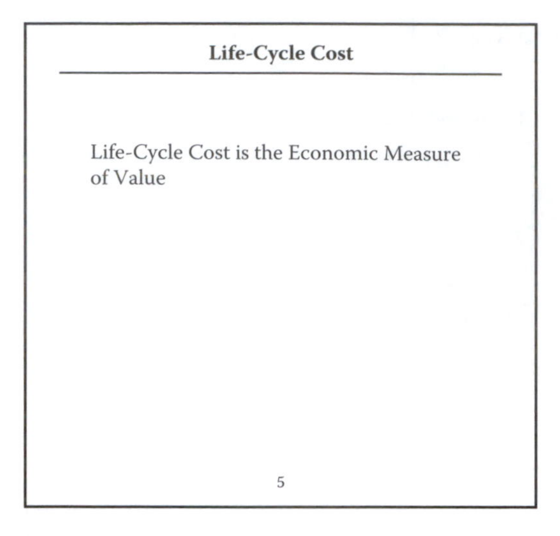


Value Analysis is not simply reduction of cost.

It is reduction of life-cycle costs. This is a combination of development cost, purchase cost, owning and operating cost, and ultimate disposal cost.

However, life-cycle cost (LCC) is *not* Value Analysis.

For example, you can choose the lowest LCC between a Cadillac, Rolls-Royce, or a Mercedes. But how did only those choices get on the list of alternatives?

Value Analysis helps you separate needs from desires—the need for transportation and the desire for a Rolls-Royce.

**What Are User Needs?**

- Availability
- Performability
- Dependability
- Durability
- Replaceability
- Maintainability
- Reliability
- Disposability

6

You satisfy all user needs—all the illities—by using life-cycle cost as your economic measure of Value.

For example, consider choice between the following two cars. You plan to drive 10,000 miles per year. You plan to keep the car for 5 years. Gas now costs $3.00 per gallon. Resale value in 5 years is 60% of what you paid for the car. Using simple constant dollar analysis, considering cost of maintenance and insurance equal, LCC indicates that to serve your transportation function, the more expensive car is the better value because it achieves better gas mileage.

| | *Car A* | *Car B* |
|---|---|---|
| | 20 mph | 30 mpg |
| Price | $25,000 | $28,000 |
| Gas | $7,500 | $5,000 |
| Resale | $(15,000) | $(16,200) |
| | $17,500 | $16,800 |

**The Same Methodology**

Value analysis
Value engineering
Value control
Value management
Value improvement
Value assurance
Value planning

7

Value Analysis began in General Electric in 1947. It was invented by Larry Miles. He worked in the GE purchasing department.

During the war, he discovered that because of shortages, he could not procure or manufacture the exact parts needed to match client and company specifications. In the spirit of getting the job done, he found that he could always achieve the functions the client wanted if given a free hand.

Since then, the methodology has been called many different names depending on its application. The Department of Defense first called it *value engineering*. For the purpose of this course, we'll call it Value Analysis because you are using Lean and Six Sigma to analyze existing products and systems.

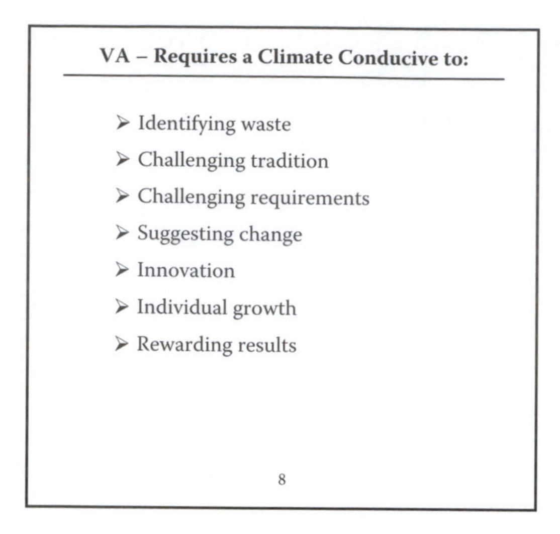


As you learned in the first week's lesson, Lean and Six Sigma require a conducive climate for change and a great deal of empathy for human relations, individual responsibility, and team work.

This requirement is no less for the effective application of Value Analysis, which is the study of FUNCTION with the objective of changing the way the function is currently being performed.

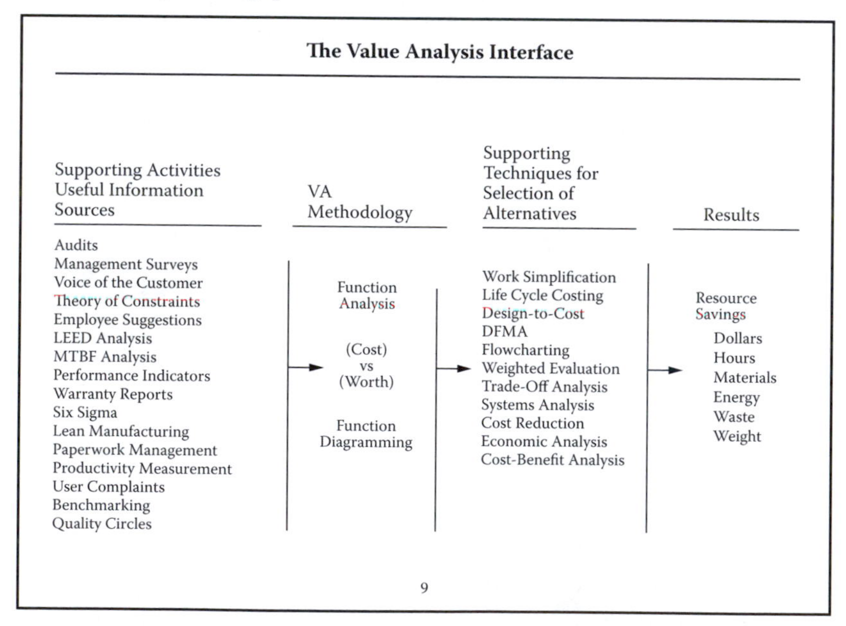


Here is a good interface picture to illustrate the close relationship among the four methodologies you are learning.

Value Analysis uses the information coming in from the Define and Measure steps you learned in the last two weeks that identify the problem (or said another way, the value improvement opportunity) to run the problem through function analysis.

Then brainstorming and other creative techniques are applied in Value Analysis to develop alternatives to performing the functions. These are then evaluated using many of the tools shown to recommend changes during the Improve and Control steps you will be learning next week.

The benefits from Value Analysis are normally expressed in life-cycle cost savings, but they can also be expressed in savings of any resource, such as time, man hours, weight, energy, and materials.

**Definitions:**
**LEED:** Leadership in Energy Efficient Design
**MTBF:** Mean Time between Failures
**DFMA:** Design for Manufacturing Assembly

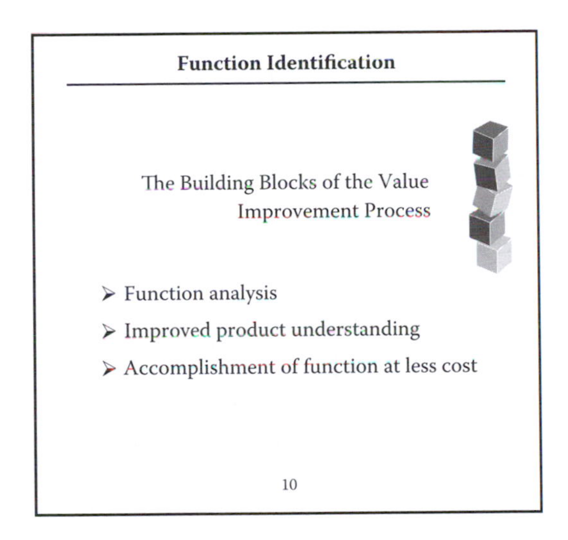


Value Analysis (VA) is the art of improving value through the analysis of a product function and its cost and worth.

VA analyzes functions to help identify unnecessary costs in a product. This procedure allows the viewing of a product objectively in terms of what it does, and must do, rather than in terms of what it currently is doing. The use of functions allows people of diverse technical backgrounds to communicate and understand each other, and the product, simply and clearly in a common language void of technical jargon.

The closest activity normally associated with Value Analysis is cost reduction; however, value analysis and cost reduction are distinctly different. Cost reduction activities are *part oriented* and the basic question asked is, "How can we make this part for less?" This usually means altering manufacturing methods, relaxing tolerances, thinning of material, and so on. Normally, this will produce savings without an alteration of the design concept.

Since Value Analysis is *function oriented*, the general question is, "How can we redesign the device to reliably accomplish the required functions for less cost?" This generally leads to new design concepts that perform needed functions more simply with higher quality and more economical manufacturing processes.

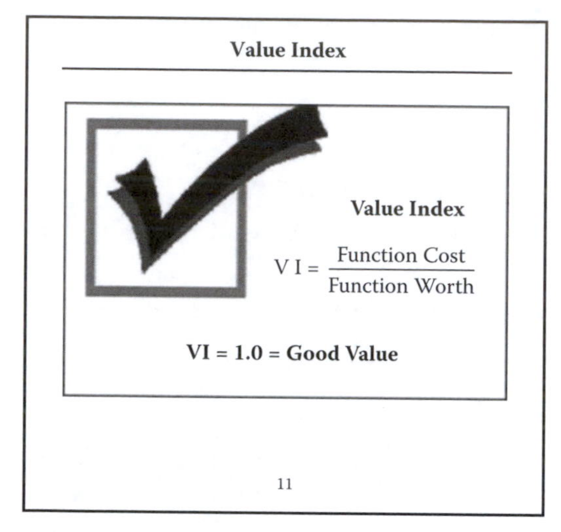


You will learn in this session how to identify function, allocate cost to function, and then determine its worth.

For example, if the cost of a function such as RECORD INFORMATION is 10 minutes by writing it on a card, but an alternate way of typing it into a computer tablet is 2 minutes, then the value index is 5.

$$\text{function cost/function worth} = 10 \text{ minutes}/2 \text{ minutes} = 5$$

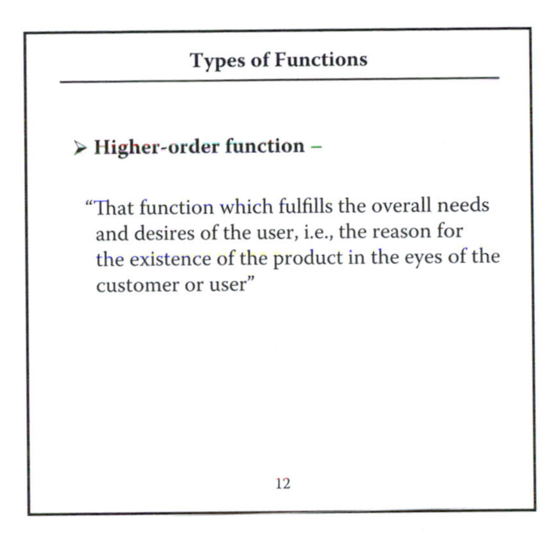


The Higher-Order Function of a product is sometimes referred to as the Task Function, Goal, or Objective of a product or process.

How it is achieved becomes the Basic Function. For example: CLOTHE PEOPLE is a higher-order function than is COLLECT CLOTHING.

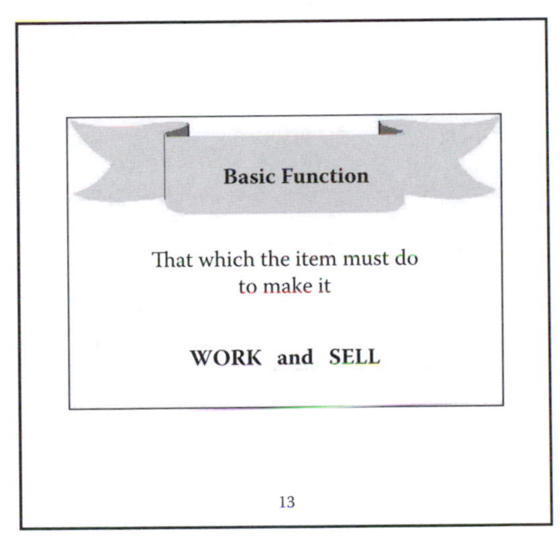


Use Value equates to a Work Function. Aesthetic/Esteem Value equates to a Sell Function.

If the product must be attractive to be obtained by the user to use it, then that function is Basic.

Aesthetics is the desire of the owner to own it! Only that amount of aesthetics required to sell the product is Basic.

Basic function normally combines both work and sell functions. For example: A refrigerator that only PRESERVES FOOD would not be wanted if it sat in your kitchen and rusted. That coat of paint and color selection is required to sell it!

**Instructor notes:** Here are some other examples:

The work function:
Screwdriver: TRANSMITS TORQUE
A light: CREATES CONTRAST
Can opener: CUTS METAL
The sell function:
Makes products marketable
The color selection in appliances
The appearance of a building

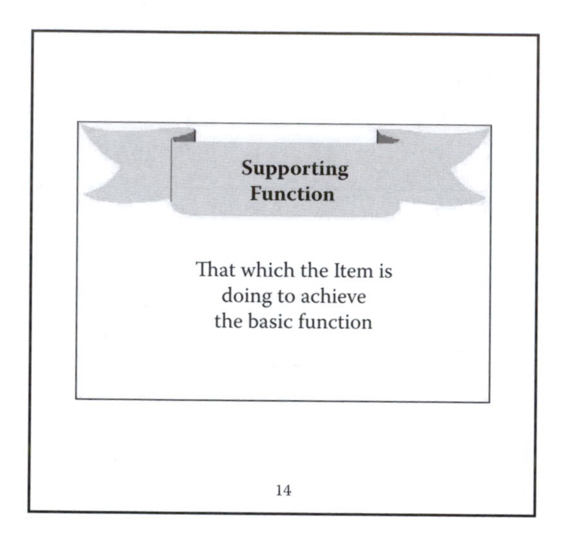


Supporting Functions are all other functions that the product performs and are subordinate to the basic function. They support the basic function and the way it is being achieved.

Some supporting functions are required and are necessary in a product or service to perform the basic function. For example, both a battery-operated flashlight and a kerosene lantern perform the basic function of producing light. A required supporting function in one, however, is to *conduct current*, while the equivalent supporting function in the other is to *conduct fluid*. Aesthetic functions that are excessive to selling the product are classified as supporting functions.

Many supporting functions are unwanted but can't easily be eliminated. In the case of the kerosene lantern, while the basic function of *producing light* is being performed, an unwanted supporting function of *produce odor* is also being performed. While an electric light is performing its basic function, an unwanted supporting function of *generate heat* is also being performed.

**Instructor notes:** Here is where aesthetic functions are supporting, not Basic.

Performance features are features that provide more capability than needed to accomplish the function, for example, gold contacts in a one-trip weather balloon.

Desires, not needs, for example, the need for a car and the desire for a Rolls-Royce; or nice to have but not necessary, for example, an ice maker in a refrigerator to preserve food.

Aesthetic functions are sensitive to art and beauty, senses—see, smell, taste, feel, hear, think, reflect, enjoy—all good verbs; and beauty—line, color, form, texture, proportion—all good nouns.

If eliminating them causes the user not to want the product, then they are Basic.

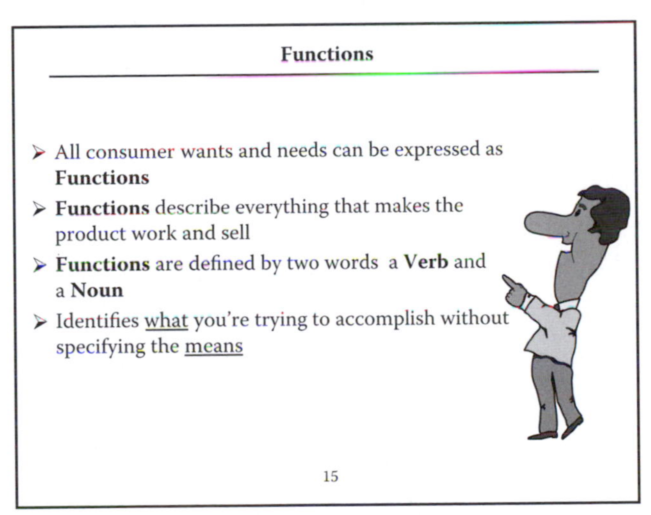


Value analysis works on everything because everything has a function!

If you can't define the function, then why spend resources on it?

You will find it very challenging to define function in two words. It is very challenging to separate needs from desires. Getting the voice of the customer will help.

For example: One might buy three pairs of the same shoes, all having the same function, but all in different colors. If the SELECT COLOR function is a need, then it too is a Basic function.

But it provides conciseness even if it takes more than one verb-noun combination to say what the product and the components of the product do.

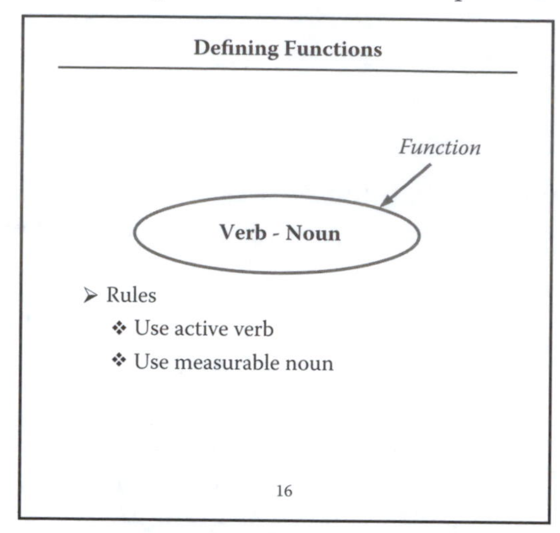


Measurable nouns—always strive for a measurable noun. This will take some time and patience. Do not use the name of a component part that is included in the project under study.

Active verbs—avoid the use of *provide*, *allow*, and *facilitate*. Often, when one of these verbs is used, the function can be reworded to become correct by making the noun a verb and finding a new noun. One should also try to avoid using verbs that end with -ize. If a passive description of a function is suspected, or you need to define a function more actively, then the "rule of thumb" is to transfer the noun you used to the verb and select another noun. For example: *provide support* becomes *support weight*.

Verbs—initially it is common to make the verb singular; that is, supports weight, resists corrosion. There is no requirement to use the verb in the singular, the plural form is acceptable.

Example: Instead of PROVIDE PUMP, more active verbs and measurable nouns would be LIFT FLUID or INCREASE PRESSURE.

**Instructor note:** Here is another example: Instead of MAKE SOUND, AMPLIFY SOUND would be better.

**Functions for Products**

| Verbs | | Nouns | |
|---|---|---|---|
| Absorb | Generate | Access | Friction |
| Access | Guide | Air | Heat |
| Actuate | Improve | Appearance | Impact |
| Apply | Increase | Bending | Light |
| Attach | Isolate | Circuit | Mass |
| Attract | Limit | Climate | Material |
| Circulate | Maintain | Cold | Moisture |
| Conduct | Pivot | Comfort | Motion |
| Connect | Position | Component | Noise |
| Contain | Prevent | Corrosion | Occupant |
| Control | Protect | Current | Parts |
| Covert | Reduce | Deflection | Path |
| Create | Regulate | Dirt | Pressure |
| Decrease | Resist | Drag | Stability |
| Direct | Rotate | Energy | Surface |
| Enclose | Seal | Entry | Torque |
| Enhance | Sense | Flow | Travel |
| Extend | Support | Fluid | Vibration |
| Generate | Transmit | Force | Weight |

17

This is only a partial list of verbs and nouns that may be used in describing the functions of a product.

The verb describes the action and the noun defines the object of that action. Searching for the most descriptive noun can be difficult and time consuming.

**Functions for Processes**

| Verbs | | Nouns | |
|---|---|---|---|
| Apply | Load | Alignment | Flash |
| Assemble | Maintain | Assembly | Gage |
| Assure | Make | Burr | Gas |
| Blend | Move | Casting | Heat |
| Clean | Position | Cause | Hole |
| Control | Prevent | Cleanliness | Inventory |
| Convert | Protect | Cold | Length |
| Create | Receive | Component | Locator |
| Decrease | Release | Container | Machine |
| Deliver | Remove | Correction | Material |
| Fasten | Repair | Damage | Mold |
| Fill | Rotate | Defect | Operation |
| Finish | Seal | Device | Part |
| Form | Store | Die | Priority |
| Identify | Supply | Dimension | Schedule |
| Improve | Thread | Dirt | Shape |
| Increase | Transport | Environment | Surface |
| Inspect | Verify | Equipment | Tool |
| Join | | Finish | Uniformity |
| | | Fixture | Waste |

18

This is only a partial list of verbs and nouns that may be used in describing the functions of a process.

**Functions for Procedures**

**Verbs**

| | |
|---|---|
| Allocate | Improve |
| Analyze | Increase |
| Audit | Inform |
| Authorize | Maintain |
| Certify | Measure |
| Compile | Monitor |
| Confirm | Obtain |
| Copy | Organize |
| Create | Procure |
| Decrease | Protect |
| Develop | Receive |
| Distribute | Reconcile |
| Enter | Record |
| Establish | Report |
| Evaluate | Set |
| Forecast | Specify |
| Generate | Test |
| Guide | Transmit |
| Identify | |

**Nouns**

| | |
|---|---|
| Alternative | Material |
| Awareness | Option |
| Concept | Order |
| Control | Part |
| Coordination | Performance |
| Creiteria | Personnel |
| Data | Plan |
| Decision | Priority |
| Design | Process |
| Deviation | Record |
| Direction | Regulation |
| Document | Request |
| Facility | Resource |
| Funds | Schedule |
| Goal | Shipment |
| History | Source |
| Information | Staff |
| Instruction | Standard |
| Inventory | Status |
| Limit | Trend |

19

This is only a partial list of verbs and nouns that may be used in describing the functions of a procedure.

**Functions for Construction**

**Verbs**

| | |
|---|---|
| Absorb | Heat |
| Alter | Illuminate |
| Amplify | Impede |
| Change | Improve |
| Circulate | Increase |
| Collect | Induce |
| Condition | Insulate |
| Conduct | Interrupt |
| Connect | Modulate |
| Contgain | Prevent |
| Conrol | Protect |
| Convey | Rectify |
| Cool | Reduce |
| Create | Reflect |
| Distribute | Repel |
| Emit | Resist |
| Enclose | Separate |
| Enjoy | Shield |
| Establish | Smell |
| Exclude | Support |
| Extinguish | Taste |
| Feel | Think |
| Filter | Transmit |
| Finish | Ventilate |

**Nouns**

| | |
|---|---|
| Air | Material |
| Appearance | Objects |
| Balance | Oxidation |
| Beauty | Parking |
| Color | People |
| Compression | Power |
| Convenience | Preparation |
| Current | Presige |
| Ego | Protection |
| Enclosure | Radiation |
| Energy | Sheer |
| Environment | Sound |
| Features | Space |
| Feeling | Structure |
| Fire | Style |
| Flow | Symmetry |
| Fluids | Temperature |
| Force | Tension |
| Form | Texture |
| Heat | Tone |
| Image | Torque |
| Landscape | Utilities |
| Light | View |
| Load | Voltage |

20

This is only a partial list of verbs and nouns that may be used in describing the functions of construction.

**Function Rules**

- **Avoid defining processes as functions:**
  - **Drill hole** is a step in a process
  - **Make opening** or
  - **Create orifice** is better!

21

One should pay close attention to and understand the differences among processes, activities, and functions. For example: *drill hole* is a step in a process; its function could be *make opening* or *create orifice.*

Another example: *file letter* is a step in a process; its function could be *store information.*

Function analysis is not easy. It is not the way people typically think about a problem.

For example: Have you ever heard anyone rushing to the grocery store to OBTAIN PROTEIN?

Changing your thought processes in this way helps generate many ideas for alternative functions. This is why Value Analysis leads to innovation. It allows you to invent on purpose, on schedule, and intentionally.

**Procedure**

- **What does the product do?**

1. Initial function determination
2. Express as: verb – noun
3. Classify as: basic (B) or supporting (S)
4. Build a FAST diagram
5. Allocate cost to function
6. Judge what the function should cost!

22

Initial function identification is the process used to brainstorm functions and place them on the function worksheet. It is the first step in defining function.

To determine the function of a product or service we ask the key question, "What does it do?" We express the answers to this question using two words—an active verb and a measurable noun.

This questioning produces a list of many functions that the item does. For example, the functions of a door might be to *seal opening, control opening*, or *permit access*. Remember, try to avoid using the verb *provide* whenever possible. It is not an active verb!

Then we classify the functions as basic or supporting (we often call these *secondary functions*).

We'll explain a FAST diagram soon.

Once all the functions are identified and classified, we assign cost and worth to them.

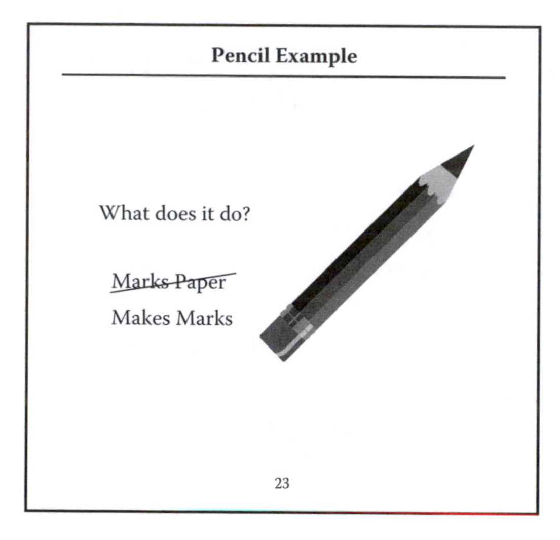


Try not to use nouns that name objects or limit the way the function is performed. Enlarge your point of view to a higher level of abstraction.

Alternative ways to MARK PAPER is more restrictive to brainstorm than is alternative ways to MAKE MARKS.

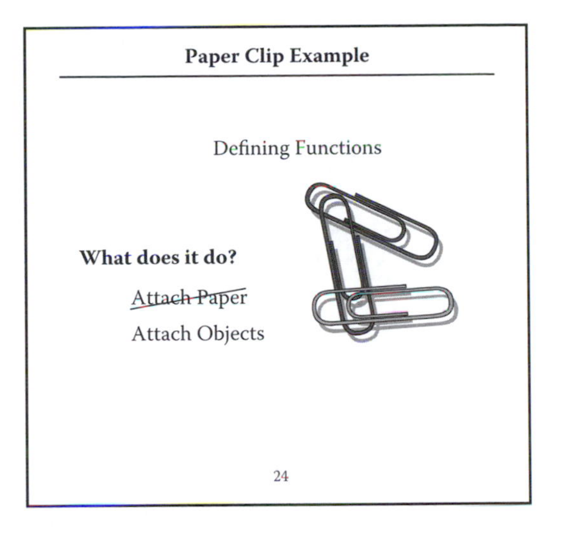


Enlarge your point of view.

*Attach objects* is a higher level of abstraction than is *attach paper*.

**Relate to User's Use**

- Basic function also must relate to user's use
- A paper clip could also be purchased and used to:
    - Pick lock
    - Clean pipe bowl
    - Ream hole
    - Etc.

25

Normally we attempt to define the intended basic function for which the product was designed.

But don't overlook the Voice of the Customer information to find out how it is being used.

Case in point: At a Naval Air Rework Facility they had to repaint an emblem on the fuselage. They had a small can of enamel on hand but could not open it. So they went to the tool crib and took out a screw driver to PRY LID and STIR PAINT!

Now the screw driver was dirty, so they threw it away. We only discovered this by observing how many screw drivers they ordered each year and wondered what happened to them. Needless to say, we purchased small sticks for them to use to STIR PAINT.

**Instructor note:** Here is an additional example to use: An electric wire could CONDUCT CURRENT or FASTEN PART depending on how it is used.

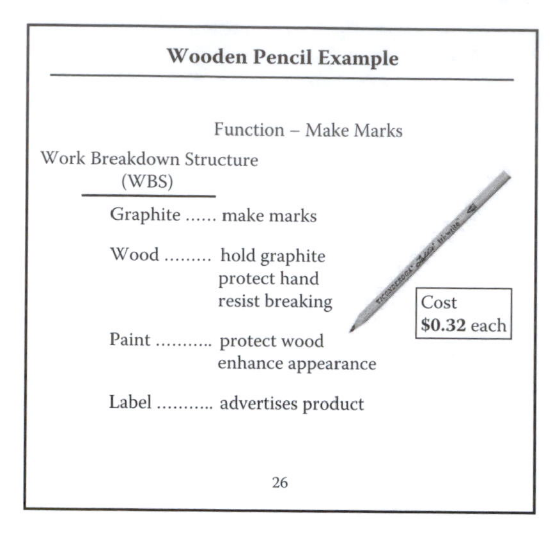


The basic function of a pencil is to MAKE MARKS.

Make a work breakdown structure (WBS) of the item. In this example it consists of four components: Graphite, Wood, Paint, and Labeling.

You will find that only the graphite performs the basic function of MAKE MARKS. All the rest of the functions are supporting functions.

The wood holds the graphite and protects the hand from getting smeared with graphite and keeps the graphite from breaking. The paint protects the wood and its color enhances appearance. The label advertises the product.

Purchased in packages of 12, each pencil costs 32 cents.

For simplicity, we have not shared the packaging component in the WBS.

In addition, we left off the eraser, which REMOVES MARKS, and the metal band that FASTENS REMOVER.

This is an example of the initial way of function identification.

**Questions for the class:** Have we identified *all* the functions? Can anyone identify any missing function?

**Instructor note:** Key points to emphasize:

Study the product as a whole first.
Study parts and pieces second.

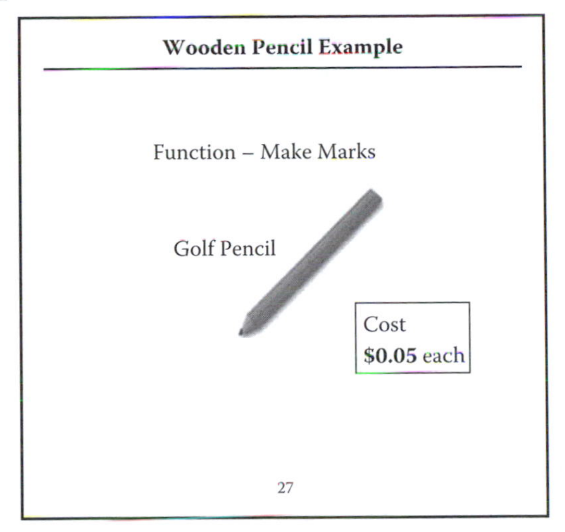


One of the ways to review for missing functions is to compare against similar products to see if there are obvious differences that create missing functions that might contain hidden cost.

The cost of these pencils is $0.05 when sold in packages of 144 each.

Why is the "golf" pencil one half the length of the previous wooden pencil, yet the cost of the golf pencil is one seventh the cost?

This implies a value mismatch that should be identified when computing the value index.

**Instructor notes:** What function is caused by the increased length of the pencil?

Verb______________ Noun______________ Answer: INCREASE LIFE

Could the golf pencil be made any shorter and reduce cost further? If not, why not? What function does the bottom half of the pencil serve that was not identified for the previous pencil?

Verb______________ Noun______________ Answer: STABILIZE PENCIL

You will find that there is a minimum length that the pencil must have to stabilize it in the hand between the fingers and the arch of the hand at the thumb.

The STABILIZE PENCIL function needs to be added as a supporting function for the Wood on the pencil. It is not required for the Graphite length because the Wood will HOLD GRAPHITE no matter what its length.

**Function Analysis System Technique (FAST)**

- Now it's time to learn about a FAST diagram
  - First, its benefit
  - Second, how to read it
  - Third, how to build one

28

There is no perfect Function Analysis System Technique (FAST) diagram solution that everyone should strive to achieve.

The only requirement is that it follows its logic diagramming rules and makes sense to you.

One last word of caution: While diagramming, STAY WITHIN THE PRODUCT YOU ARE STUDYING! This will be a very difficult task because FAST is a very strong design tool for making new products and systems.

Asking the question, "What does it do?" is staying within the product.

Asking the question, "What should it do?" is designing something new.

**Benefits of FAST**

- Identifies missing functions from the initial selection technique
- Identifies the basic function and the higher-order function
- Groups functions that happen all the time and functions that are criteria
- Establishes the scope of the project study
- Provides a good presentation tool

29

The act of making a FAST diagram will really help the team understand the problem.

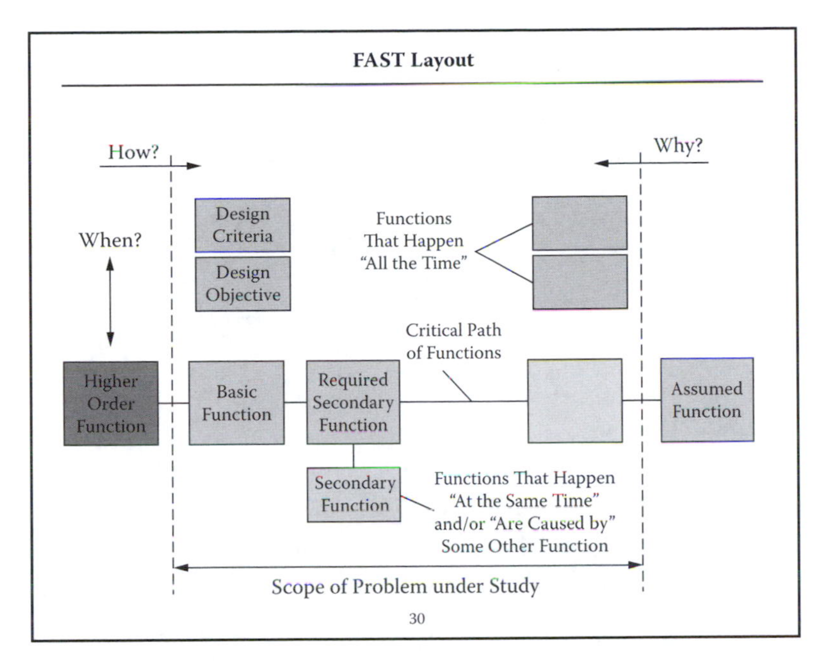


The original FAST diagramming was called Technical FAST. There is also a Customer FAST. We will be using a blend of both in this course.

Here is an overview of how to read the diagram:

- Basic function and higher-order function are separated as shown by the left scope line.
- Assumed functions are those given at the start of the study and are set just to the right of the right scope line so that they are not within the scope of the study.
- The critical path is the line of functions between the basic function and the assumed functions.
- Time is displayed vertically in the diagram, so functions that happen concurrently (at the same time) are placed above one another.
- Criteria functions and functions that happen all the time are placed into storage areas out of the stream of cause and effect. For example, the markings on a pencil, the function of IDENTIFY MANUFACTURER, occurs all the time. The function RESIST BREAKING might be a design criteria.

**Instructor note:** Background information: Sometimes viewed as a horizontal "ladder of abstraction." More abstract functions are successively to the left, less abstract to the right.

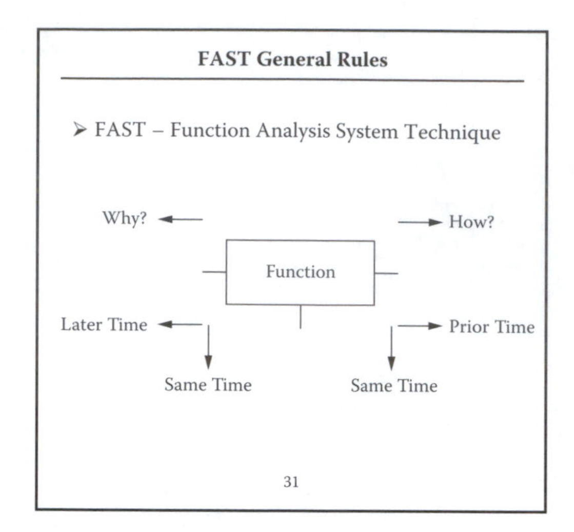


In the early years of value analysis we tended to generate functions randomly, and to start a FAST we still do. However, if this is the only way to identify functions, it will miss a few important functions.

FAST diagramming was developed to help us do a more thorough job of function identification. FAST is just a tool. The process of making the diagram by the team is important, not the perfection of the resulting diagram.

Take any function and ask "Why?" and the answer should be a function to its left. In our door example, ask "Why CONTROL OPENING?" The answer to the left might be "PERMIT ACCESS and SEAL OPENING."

Take the same function and ask "How?" The answer should be placed to the right. How do we CONTROL OPENING? The answer might be "INSTALL DOOR!"

If two functions happen at the same time, they are positioned above one another.

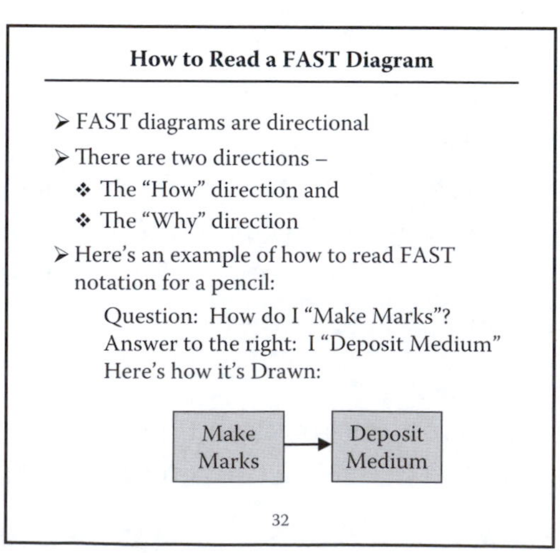

The "Why" question is answered to the left. The "How question is answered to the right.

Be careful while you are diagramming. Asking "Why?" and "How?" will lead you into challenging requirements.

If you come up with ideas, just write them down for later use in your study. Right now, stick to the task of performing function analysis on the product at hand, just as it was designed to do.

**Mouse Trap Example: Use the Why–How Grid**

- Place identified functions down the center of the function worksheet
- Fill in answers of questioning why (function) to the left
- Fill in answers of questioning how (function) to the right

| Function Worksheet | | |
|---|---|---|
| Why? | Function | How? |
| | | |
| Kill Mice | Apply Force | Store Torque |
| | | |
| | | |
| | | |
| | | |

33

The Why–How grid is used to press the search for missing functions not initially identified.

Place one of your initial verb-noun functions in the center, such as APPLY FORCE. Ask why this function is performed and place the answer in the left column in verb-noun form such as KILL MICE. This will increase your level of abstraction.

Now ask how the center function, APPLY FORCE, is performed and place the answer in the right column, such as STORE TORQUE. This will be more specific and reduce your level of abstraction.

Do this at least once for all functions you have initially identified.

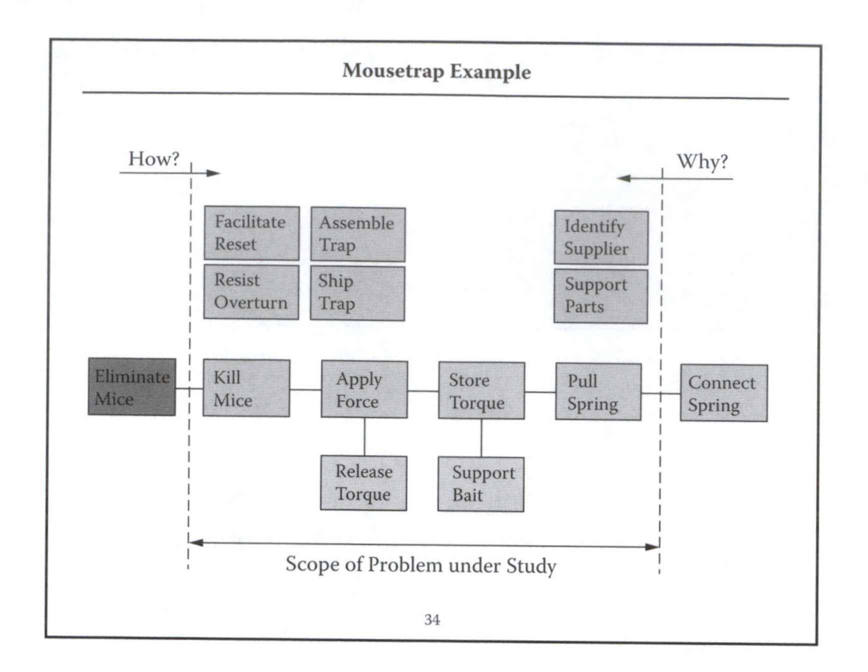


**Instructor notes:** Here is a FAST diagram of a mousetrap.

Its basic function is KILL MICE! How? By APPLYING FORCE at the same time you are RELEASING TORQUE to hit the little buggers!

Have the class quickly brainstorm alternate ways to KILL MICE.

Typical answers: poison, shooting, stabbing, stepping on them, and so on.

Why do you PULL STRING? The answer is to the left—to STORE TORQUE. Notice at the same time that you are SUPPORTING BAIT. Why STORE TORQUE? The answer is to the left—to APPLY FORCE.

If the left scope line were moved one box more to the left, the basic function would be ELIMINATE MICE. That would open up a whole list of options such as moving away from them to eliminate them or even sterilizing them rather than killing them.

Next, have the class brainstorm ways to ELIMINATE MICE.

Typical answers: move away, keep a cat, sterilize them, keep a clean house, and so on.

The scope line is important since it defines the problem being studied. Moving the left scope line to the left enlarges your point of view to a higher level of abstraction. It is only productive to enlarge your point of view if you think you can implement a solution. If you are in the mousetrap business, eliminating mousetraps as a way of eliminating mice might put you out of business, or you might stumble on a different product to manufacture. Notice that *the* out of scope assumed function is the fact that you are studying mousetraps so you are accepting that you will CONNECT SPRING.

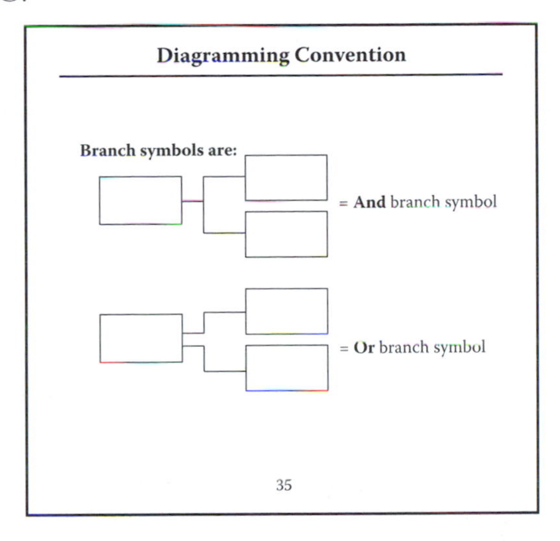


It is possible to have a SPLIT or BRANCH on the CRITICAL PATH, that is, having 2 (or more) functions leading to a preceding and/or a succeeding function.

You may have a major function logic path and a minor function logic path.

**Minor Function Logic Path:** Branches can also spawn from a function that is caused by or happens at the same time as a critical path function; thus, it is either above, or more commonly below, the critical path. In some cases a new How–Why logic path is started from this branch function, creating a minor function logic path.

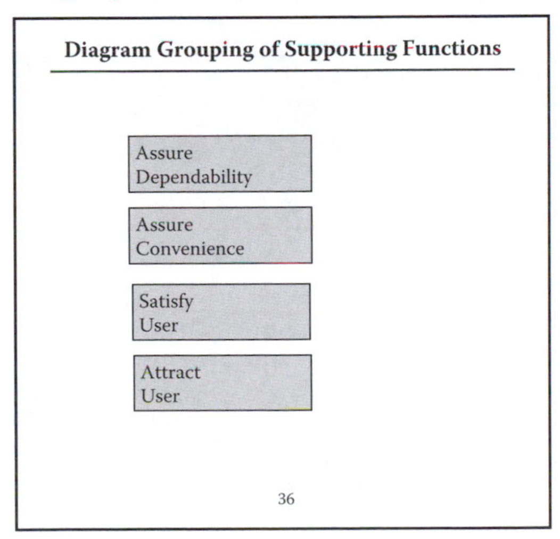

One can group primary Supporting Functions into one of the following four groups. Many of these functions can be identified from Voice of the Customer input:

- *Assure Dependability:* Those functions that tend to minimize deterioration, that is, add strength, corrosion protection, protection of persons or environment, and so on.
- *Assure Convenience:* Those functions that make the product or process easier to use, that is, instructions, aids to servicing, cleaning, repairing, correcting, and so on.
- *Satisfy User:* Those functions that enhance the product or process above customary expectations, that is, (smaller, faster, lighter), physical comfort, status, and so on.
- *Attract User:* Functions that appeal to the senses, that is, noise level, appearance, performance, speed, strength, and so on.

**Special Care Rules for FAST**

- Maintain a constant frame-of-reference. Don't mix functions performed by the user, designer, or manufacturer.
- Avoid using as a noun the name of a part, labor operation, or activity.

37

**Steps to Create a FAST Diagram**

1. Define functions initially
2. Identify missing and higher-order functions
3. Group functions
4. Identify basic functions and start the critical path
5. Add the 4 standard supporting functions
6. Expand diagram
7. Verify diagram

38

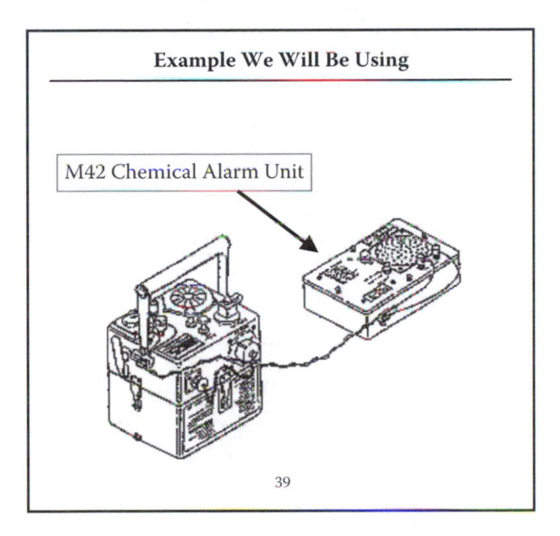


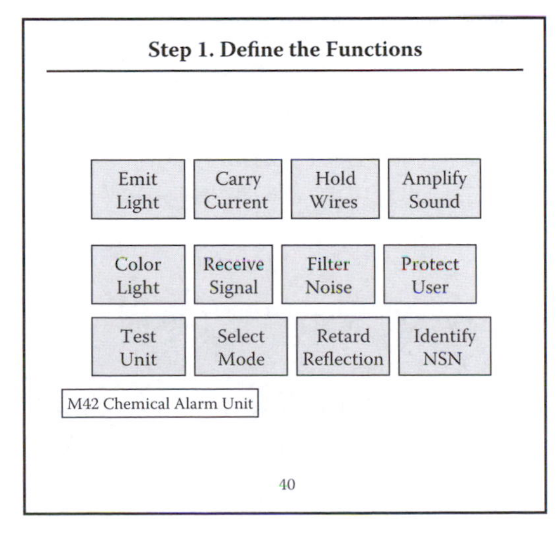

Here are twelve randomly identified functions of the M42 Chemical Alarm Unit. To get these functions we asked, "What does it do?"

Using the verb-noun method, define functions performed by the product, or by any of the parts or labor operations of the product.

If this were an actual problem, each of these initially identified functions would be written on Post-it® Notes.

We are sure that there are some missing functions. Next we'll start to see how to tell what they are.

**Instructor note:** Have the students make a blank Why–How grid on a piece of paper by drawing two vertical lines creating three columns on the paper.

Have them write AMPLIFY SOUND in the center (function) column as a good place to start because it looks like a Basic function.

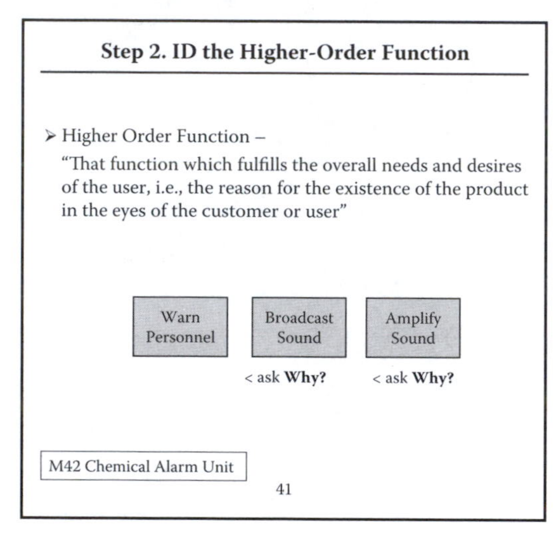


From the previous list of initially generated functions, select the one that you think is Basic.

In our example, the only one on the previous list that is nearest to representing the Basic function is AMPLIFY SOUND.

Ask, "Why AMPLIFYING SOUND?" The answer is placed to the left—to BROADCAST SOUND.

Next, ask "Why BROADCAST SOUND?" The answer is again placed to the left—to WARN PERSONNEL.

You have now just added two functions to the list for this product and have identified the higher-order function for the product as WARN PERSONNEL.

You should iteratively use the Why–How grid on all of the initially generated functions. The list will expand rapidly. In practice, Post-it Notes would be created for all new functions.

**Instructor note:** While you are briefing the slide, tell students to write BROADCAST SOUND to the left of AMPLIFY SOUND (in the Why column).

Then write BROADCAST SOUND under AMPLIFY SOUND in the center column and ask "Why?" again, and put WARN PERSONNEL in the Why column.

**Step 3. Group Functions**

- Group functions as follows:
  - Basic Function: This function is essential to the performance of the higher-order function.
  - Criteria or functions that "happen all the time."
  - Supporting function: A function supporting the method of achieving the basic function or critical for user acceptance of the product.

42

There are three groups to make:

1. Pick out what you think are the Basic functions and ask if the function can be eliminated and still satisfy the user. If the answer is NO, the function is Basic.
2. Select functions that seem to be criteria or will "happen all the time."
3. The balance of the functions should be supporting functions.

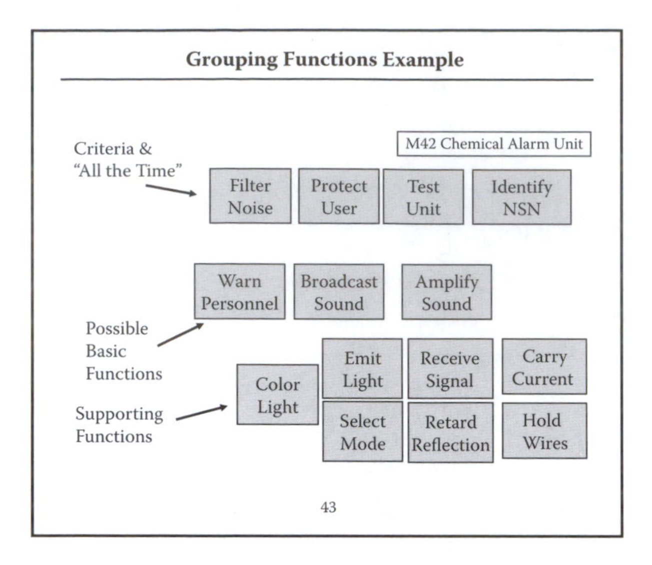


We have grouped all the functions from initial analysis, the Why–How grid, and constructed in the previous step.

The higher-order function and basic function start our critical path.

The Criteria and all the Time functions are moved to temporary storage at the top of the sheet.

The remaining Supporting functions fall in the bottom group.

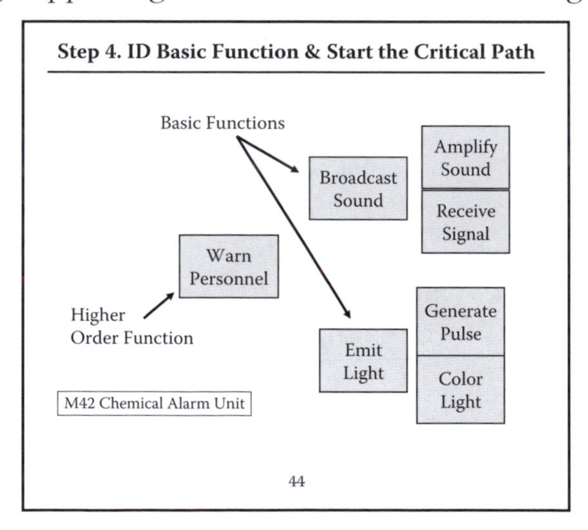


Further analysis shows that personnel are warned by this system in two fashions.

How are they warned? By BROADCAST SOUND and EMITTING LIGHT. Thus, both functions are on the critical path. If neither function occurs, then none of the personnel are considered to be warned—including the blind and the deaf. That makes those two functions Basic!

How each is achieved are the functions shown to the respective right of each.

**Instructor note:** Insure that the students realize why EMIT LIGHT is a Basic function not to be overlooked.

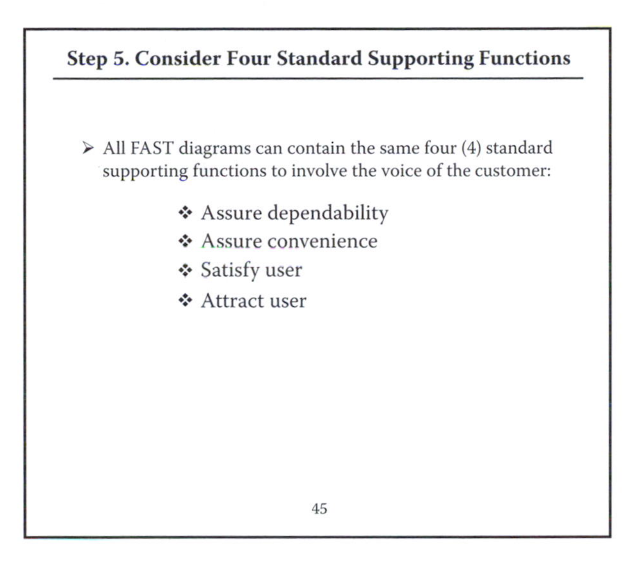


When you have Voice of the Customer input, try to express their desires in verb-noun function form.

Group all supporting functions into the four categories (shown in the figure) in which they belong by moving each of the cards to the right of its primary supporting function.

Next, assemble a tree-structure for each of the primary supporting functions. To do this, determine if there are two or more supporting functions that fit into logical "groups."

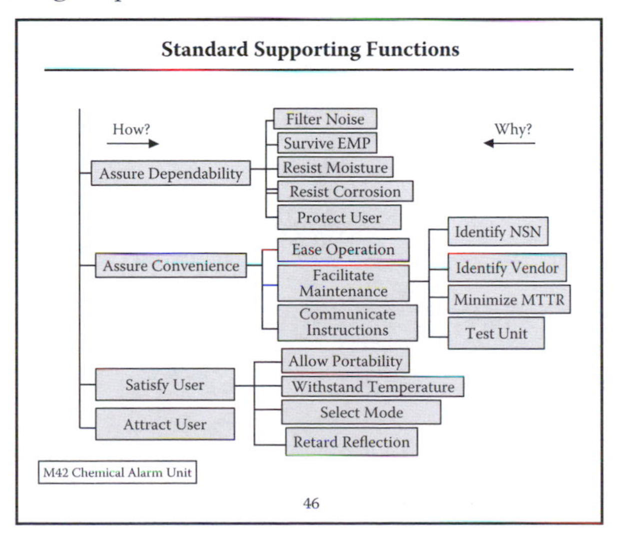

As you can see, once you add these four standard supporting functions and ask *how* each of them is achieved, many more functions come to light.

In fact, many of the Criteria functions and functions that happen all the time will fall into place under these four standard functions.

**Step 6. Expand FAST Diagram to the Right**

- First, perform function expansion on the basic function structure, then on the supporting function structure.
- Next, create a logically valid tree-structured diagram which directly complies with the **How** rule, and involves no bridging or conditioning phrases, involves lots of thought and discussion.

47

Keep asking, "How does it (the product) do this?"

Most answers will be found among the existing functions. Add functions as needed.

**Note:** Be sure to constantly maintain the view point of a *user* as you go through this questioning.

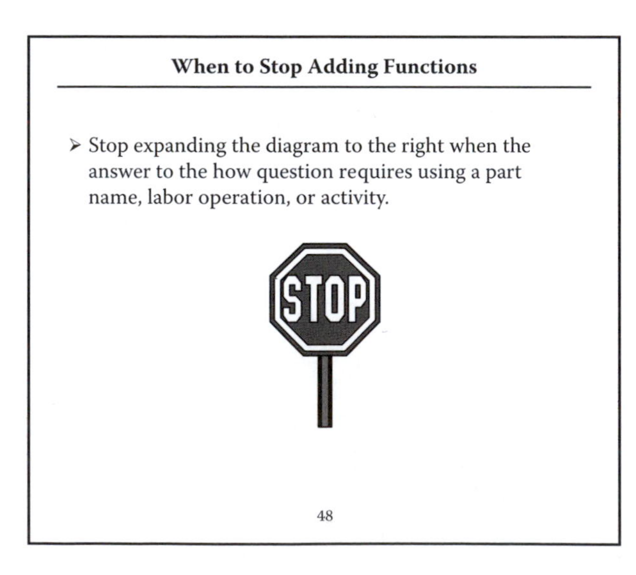

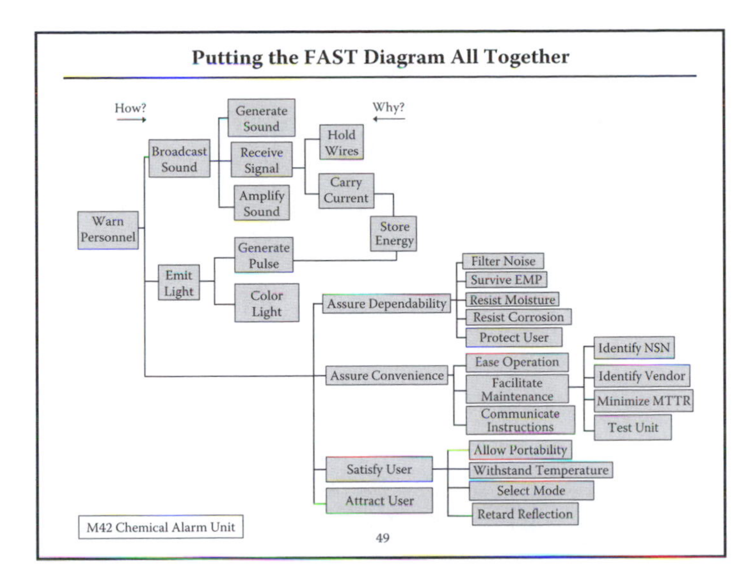


We will set back the four standard supporting functions, identified in figure 46, a bit because none of them are Basic functions.

The left scope line has not yet been set. You can either study alternate ways to WARN PERSONNEL (those ideas would definitely do away with the M42 Chemical Alarm Unit), or you can study alternative ways to simultaneously BROADCAST SOUND and EMIT LIGHT.

The right scope line has not yet been set. The right scope line will be set depending on whether it is decided to include STORE ENERGY in the analysis.

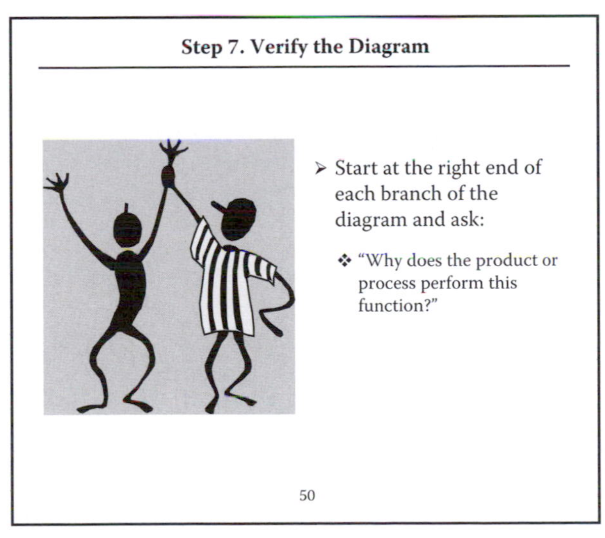


Last, the whole team should read the diagram out loud from right to left and then left to right.

If everyone feels it is logical and tells the story, you are finished!

However, don't be afraid to tweak it to add a missing function that would clarify something.

It is very rare than separate teams will create exactly the same diagram. Remember, we said in the beginning that there is no pat answer—just a logical function expression of what is happening.

# Appendix D: Product-Oriented VE Training Material for LSS Black Belt Training

**Function – Cost-Worth**

➢ Steps

1. Allocating cost to function
2. Determining worth of function
3. Calculating the value index

1

Now that you have made the Function Analysis System Technique (FAST) diagram and determined all the functions, it is time to find out (if you haven't already) what each function costs and what that function is worth.

- Look at function costs
- Identify value mismatches

A value mismatch is where it is apparent that cost does not equal worth. Using this information to calculate the value index for each function should help to prioritize what improvement would be most beneficial.

**Allocating Cost to Function**

- Allocate product cost to components
- Allocate component cost to its functions
- Pro-rate the cost when more than one function is being performed.
- Isolate function cost by comparing similar products

**Best Judgment – Not Accuracy**

2

Isolate each function. For Example: A solid core 3'0" × 6'8" interior door (paneled type) might cost $290.00. It provides three functions (assumed): SEALS OPENING, DEADENS SOUND, LOOKS ATTRACTIVE.

To prorate the cost to each of these three functions, isolate them and determine the cost of comparable doors that do not serve the other two functions.

- If a plain hollow-core door costs $172.00, then $172.00 is a good cost for the function SEALS OPENING.
- If a plain solid-core door costs $214.00, then $42.00 is a good cost judgment for the function DEADEN SOUND.
- The balance of the cost, $76.00, between a plain solid-core door, and a paneled solid-core door, must go for the function LOOK ATTRACTIVE.

| *Function* | *Cost* |
|---|---|
| Seal opening | $172.00 |
| Deaden sound | 42.00 |
| Look attractive | 76.00 |
| **Total** | **$290.00** |

Using best judgment to obtain reasonable cost allocation on a rational basis is more important than explicit accuracy.

**Instructor note:** Elaborate with the following:

- Cost of a product to the user is its Price.
- The sum of the costs of the components of a product should equal the product's cost.

**Worth**

- ➢ The lowest cost to achieve a function
- ➢ Determine by:
  - ❖ Comparison
  - ❖ Historical data
  - ❖ Personal experience

3

*Worth* is just a technique, not an absolute value.

- It is based upon the evaluator's judgment and experience.
- It is used only as a tool to identify the value index relationships of functions.

It would be presumptuous to attempt to tell someone the worth of a product. However, it is not presumptuous to tell them the worth of a function.

The lowest-cost method that you can think of to achieve a function is used to indicate the worth of that function.

**Determining Worth of Function**

➢ Process to determine worth

1. Select the individual function
2. Creatively think of other ways that function has been performed
3. Associate or assign approximate cost to it
4. Add up values of several functions to get total worth
5. Calculate the value indexes

4

Worth is determined by developing or thinking of other methods of performing functions.

Example: What is it worth to "Open or Close circuit"? A light switch that performs this function would cost $10.00. A lower-cost item or method that performs this function would have a lower worth.

The (open/close circuit) function is the basic function of a manual fire alarm box, which costs about $150.00.

Where an item has several functions, you must determine the worth of each function separately and add them together to get overall worth to compare with overall cost.

Example: In the previous door example, the worth of SEAL OPENING might be $25.00 if it were done using gypsum board.

DEADEN SOUND might be worth $10.00 if insulation were placed between the sheets of gypsum board.

LOOK ATTRACTIVE might be worth $50.00 if both sides were laminated with paneling.

Therefore, the overall worth of the door's functions is $85.00. However, the overall cost of the doors functions is $290.00. Its value index = $290/$85 = 3.4.

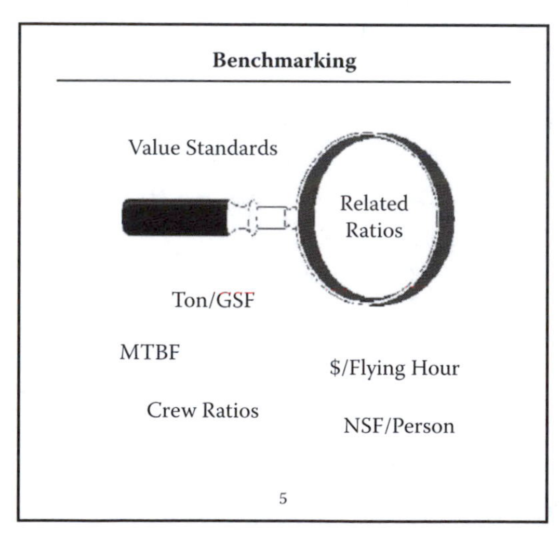


Value standards developed by experienced organizations can also be used as indicators of overall worth. These standards are also called *benchmarks* since they reveal a goal that others have achieved.

In the building industry, for example, you can compare the tons of air conditioning provided per gross square foot (GSF) with a reasonable standard of 300 square feet per ton to spot value mismatches.

In manufacturing, the mean time between failures (MTBF) for a product can be set as a standard and often influences warranty periods.

Efficiency of the use of space is obtained by dividing net square feet (NSF) by gross square feet. For commercial space, good value normally is greater than 80%. Government space, however, normally runs at 70% efficiency. If your building design is worse than that, you have another value mismatch.

Benchmarks and standards represent achievable goals, not necessarily the lowest cost to perform a function. However, if that is all you have to compare to, it is better to settle for that than nothing at all.

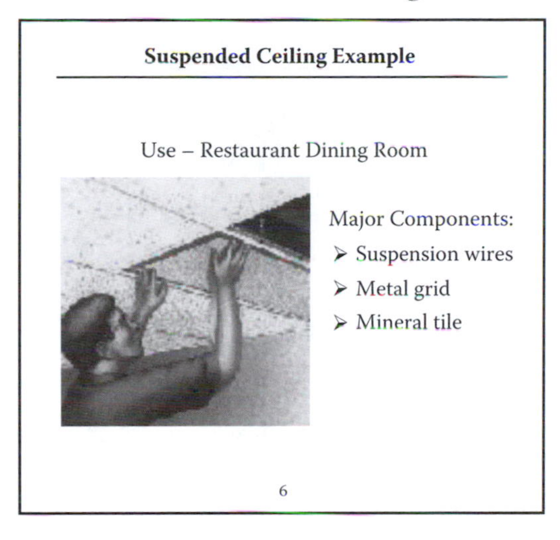


Here is an example of a standard suspended ceiling used in a restaurant dining room indicating the procedure to follow in performing function–cost–worth analysis.

**Allocate Cost to WBS Components**

Function Analysis Worksheet
Project: Restaurant Dining Room Ceiling

| Item or Component | Verb | Noun | Kind | Function Cost $/SF |
|---|---|---|---|---|
| Suspended Ceiling | | | | $ 4.34 |
| Wire Hangars - #9 gage | | | | $ 0.17 |
| T-bar System - 2×4 grid | | | | $ 1.39 |
| Mineral Fiber Tile - 5/8" | | | | $ 2.78 |

7

Using the Function Analysis Worksheet.

- List the components of the item under study (this is like the work breakdown structure).
- List the cost of each component.

The total cost of the product on the top line (the ceiling, in the example shown) must equal the sum of its components.

For this example the RSMeans cost guide was used to cost these components.

**Add Functions to WBS Components**

Project: Restaurant Dining Room Ceiling

| Item or Component | Verb | Noun | Kind | Function Cost $/SF |
|---|---|---|---|---|
| Suspended Ceiling | | | | $ 4.37 |
| | Hide | Structure | B | $ 4.07 |
| | Improve | Appearance | B | $ 0.30 |
| Wire Hangars - #9 gage | | | | $ 0.19 |
| | Support | Weight | S | $ 0.08 |
| | Determine | Distance | S | $ 0.08 |
| | Retard | Corrosion | S | $ 0.03 |
| T-bar System - 2×4 grid | | | | $ 1.40 |
| | Support | Weight | S | $ 1.15 |
| | Close | Cracks | S | $ 0.10 |
| | Retain | Materials | S | $ 0.15 |
| Mineral Fiber Tile - 5/8" | | | | $ 2.78 |
| | Hide | Structure | B | $ 2.25 |
| (Fissured Type) | Improve | Appearance | B | $ 0.30 |
| | Retard | Fire | S | $ 0.23 |

8

The next step is to add the Functions to the Function Analysis Worksheet and then allocate a component cost to each function.

The units of cost used should be the same. In this case, cost is based on cost/square foot of ceiling covered by the suspended ceiling system.

This is not an exact science. You do this by experience and judgment. You do it by comparison of cost of similar products that are with or without the functional feature being priced.

Consider these examples:

- You can get mineral fiber tile without the RETARD FIRE feature for $0.23 less; therefore, it can be assumed to be the cost of that function.
- You can get plain flat mineral fiber tile for $0.30 less than the cost of the fissured texture type. Therefore, the cost of the IMPROVE APPEARANCE function can be assumed to be $0.30.

- Adding these up = \$0.53. This makes the cost of the HIDE STRUCTURE function \$2.25. All three components add up to \$2.78, which is the whole cost of the 5/8" mineral fiber tile component.

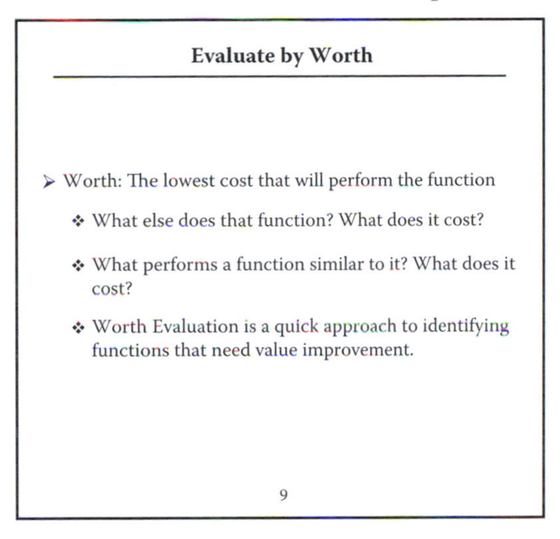


Think of the lowest cost, cheapest way that you know of that would serve the function. Without regard to failure, that lowest cost is a comparative worth of the function.

**Determining Function Worth**

Project: Restaurant Dining Room Ceiling

| Item or Component | Verb | Noun | Kind | Function Cost \$/SF | Tenatative Alternate | Function Worth \$/SF | Value Index |
|---|---|---|---|---|---|---|---|
| Suspended Ceiling | | | | \$ 4.37 | | \$ 2.18 | 2.00 |
| | Hide | Structure | B | \$ 4.07 | Cloth Drape | \$ 1.88 | 2.16 |
| | Improve | Appearance | B | \$ 0.30 | Color | \$ 0.30 | 1.00 |
| Wire Hangars - #9 gage | | | | \$ 0.19 | | \$ 0.16 | 1.19 |
| | Support | Weight | S | \$ 0.08 | Horizontal Wire | \$ 0.05 | 1.60 |
| | Determine | Distance | S | \$ 0.08 | | \$ 0.08 | 1.00 |
| | Retard | Corrosion | S | \$ 0.03 | Stainless Steel | \$ 0.03 | 1.00 |
| T-bar System - 2×4 grid | | | | \$ 1.40 | | \$ 0.12 | 11.67 < |
| | Support | Weight | S | \$ 1.15 | Parralel Wire | \$ 0.10 | 11.50 |
| | Close | Cracks | S | \$ 0.10 | Pin Cloth Edges | \$ 0.02 | 5.00 |
| | Retain | Materials | S | \$ 0.15 | Drops Out | \$ - | 0.00 |
| Mineral Fiber Tile - 5/8" | | | | \$ 2.78 | | \$ 1.90 | 1.46 |
| | Hide | Structure | B | \$ 2.25 | Cloth | \$ 1.50 | 1.50 |
| (Fissured Type) | Improve | Appearance | B | \$ 0.30 | Color | \$ 0.30 | 1.00 |
| | Retard | Fire | S | \$ 0.23 | Flame Retardant | \$ 0.10 | 2.30 |

10

Complete the Function Analysis Worksheet by brainstorming tentative alternates to satisfy each function. The cost of those alternatives becomes *function worth.*

Remember, worth is the lowest-cost way you can think of to achieve a function in isolation of all other functions and without considering the consequences of failure.

As cost in the work breakdown structure (WBS), worth of functions adds up to the worth shown for the component that performs those functions.

The value index is simply achieved by dividing function cost by function worth.

Where a function has zero worth, just make the index 0 or leave it blank for simplicity.

In this example, the SUPPORT WEIGHT function shown by the wire hangars and T-bar system has the highest index and is ripe for ideas to improve value.

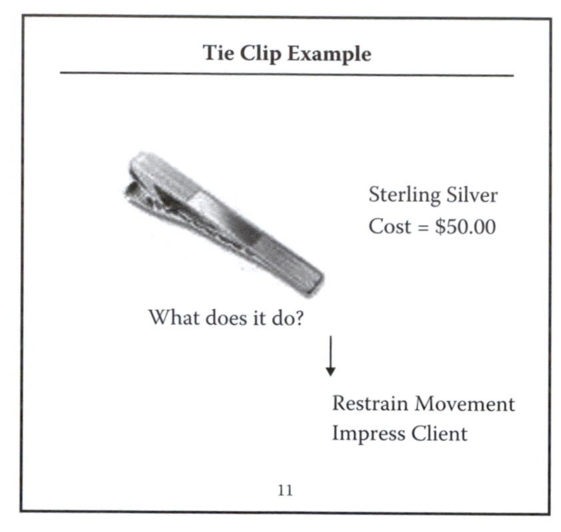


This is an example using a tie clip. First, collect all the information you can about it. What is it made of? What does it cost? What does it do?

An easy way to determine function of an item is to put yourself in the place of the item. Visualize, "I am a tie clip. What do I do?" Or, "Why do I wear a tie clip?"

**Value Index Calculation**

| Item or Component | Verb | Noun | Kind | Function Cost | Tentative Alternate | Function Worth | Value Index |
|---|---|---|---|---|---|---|---|
| Tie Clip | | | | | | | |
| Steel Shape | Restrict | Movement | B | $ 20.00 | Paper Clip | $ 0.01 | 2,000 |
| Sterling Silver | Impress | Client | B | $ 30.00 | Coating | $ 0.04 | 750 |
| | | | | $ 50.00 | | $ 0.05 | 1,000 |

12

The calculation of a value index for the tie clip is done by dividing function cost by function worth.

First, allocate cost to function. The cost of the tie clip ($50) is allocated by judgment to its two functions. A plain tie clip that would RESTRICT MOTION could be obtained for just $20. Making it of sterling silver to IMPRESS CLIENT costs another $30.

Second, determine the worth of each function. A quick comparison for the lowest-cost item that one could think of that would just RESTRICT MOTION is a paper clip, whose function worth is just one cent. Coating paper clips so they would look like gold or silver to IMPRESS CLIENT would make that function worth another four cents.

An index of 1 indicates good value. The overall index of 1,000 indicates that there is much room for value improvement. The index of 2,000 indicates that one should start the value study by brainstorming alternative ways to RESTRICT MOVEMENT.

Now brainstorm how many other ways you can think of to RESTRICT MOVEMENT.

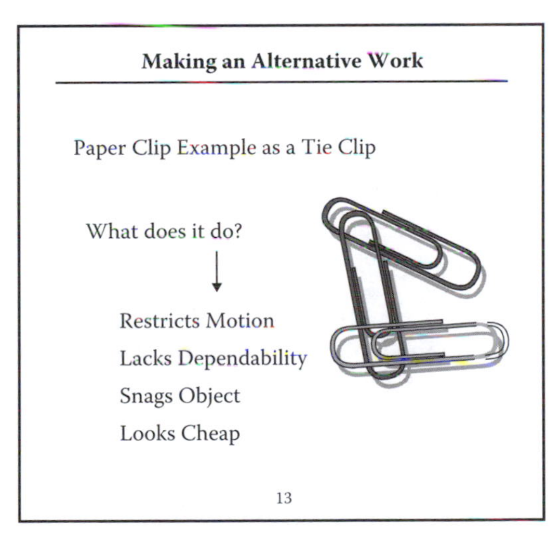


The paper clip does restrict motion, but it also has some unwanted functions:

- It LACKS DEPENDABILITY because it is too short to span the tie and eventually falls off.
- It SNAGS OBJECTS, especially wool ties because its points are too sharp.
- It LOOKS CHEAP and fails to IMPRESS CLIENTS because everyone knows it is just $0.01.

So in the spirit of value improvement, let us overcome these roadblocks by adding some value back into the product. This is an example of incrementally ADDING VALUE back into something to overcome objections and satisfy users.

We can make it longer to make it more dependable, we can make it out of smooth plastic to reduce its snagging, and we can add color and form to make it impressive.

**4.5 inch Clip**

| Item or Component | Verb | Noun | Kind | Function Cost |
|---|---|---|---|---|
| Plastic Clip | | | | |
| Spring Shape | Restrict | Movement | B | $ 0.10 |
| Plastic | Resist | Snagging | S | $ 0.10 |
| | Resist | Corrosion | S | $ 0.05 |
| 4.5" size | Impress | Client | B | $ 0.25 |
| | | | | $ 0.50 |

14

As a sample idea, here is a clip to RESTRICT MOVEMENT that costs just 50 cents.

The fact that it is plastic adds two secondary functions. It will RESIST CORROSION like a plain old steel clip might not do. It will RESIST SNAGGING on the tie like a small clip with sharp ends might do.

If you've never seen one this size it will IMPRESS you. In fact, if you were a value analyst and wore one to work people might think you were good at your job!

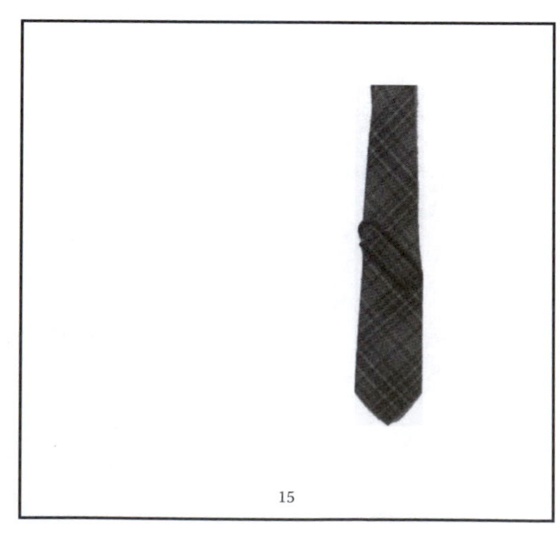

15

Is this good value? No, it is just *better* value.

Good value would be not using a tie clip at all to RESTRICT MOVEMENT and sticking the tie into your shirt.

Another way would be taking off the tie and placing it on a table with a book on top of it. Now see if it MOVES!

Remember, the user determines value, not the manufacturer or designer.

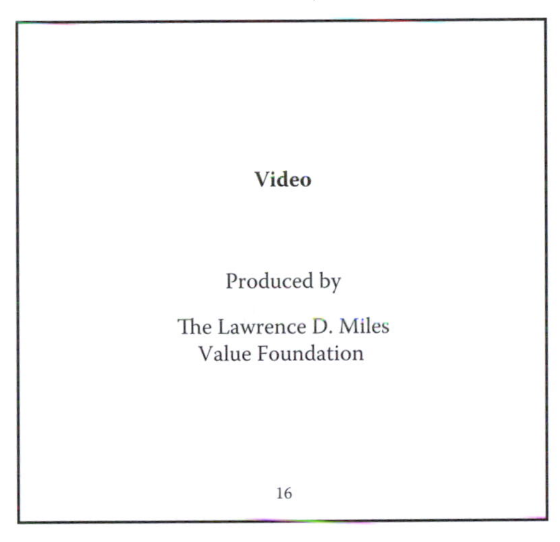


SAVE International, at http://www.value-eng.org/value_engineering.php, is the value analysis (or value engineering) professional society.

SAVE is the distributor of a Lawrence D. Miles Value Foundation DVD entitled *Principles of Value Analysis/Value Engineering*. It is a 34-minute presentation to introduce the concepts and benefits of value analysis. Topics covered include the definition of function, application of function analysis system technique, job plan, and the use of team dynamics.

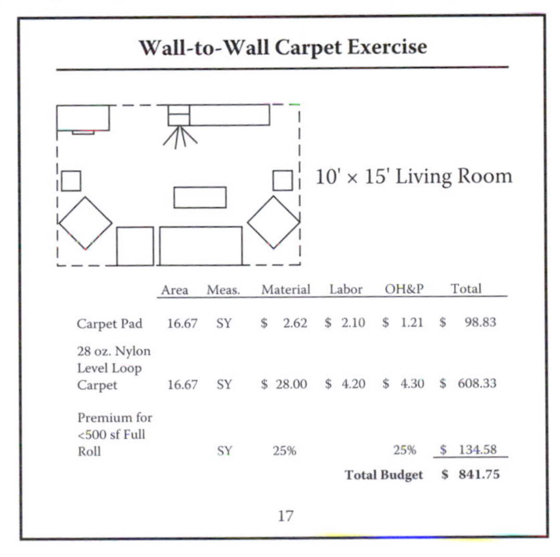


| | Area | Meas. | Material | Labor | OH&P | Total |
|---|---|---|---|---|---|---|
| Carpet Pad | 16.67 | SY | $ 2.62 | $ 2.10 | $ 1.21 | $ 98.83 |
| 28 oz. Nylon Level Loop Carpet | 16.67 | SY | $ 28.00 | $ 4.20 | $ 4.30 | $ 608.33 |
| Premium for <500 sf Full Roll | | SY | 25% | | 25% | $ 134.58 |
| | | | | | **Total Budget** | **$ 841.75** |

Here is the situation: You are selling an old house and moving into a new house. The new home has beautiful hardwood floors, but you realize that with four kids running around on wooden floors, you need carpet to reduce the sound of foot steps.

New wall-to-wall carpeting in the new house will cost almost $842 just for the living room. It would cost too much if you had to also do the other rooms as well.

**Step 1 – Function Determination**

- Place yourself in the position of the carpet.
- Ask the question "If I were a carpet what would I do?"
- Brainstorm and list probable functions for this carpet and the user's use of it!

18

**Instructor note:** Ask the students to respond with active verbs and measureable nouns such as the following:

- Minimize static
- Protect surface
- Reduce slipping
- Insulate floor
- Add color
- Reduce noise

**Step 2. Use the Why–How Grid**

Place your functions down the center of the function worksheet.

Fill in answers of questioning Why (function) to the left

Fill in answers of questioning How (function) to the right

| Function Worksheet | | |
|---|---|---|
| Why? | Function | How? |
| | | |
| | | |
| | | |
| | | |
| | | |
| | | |

19

**Instructor note:** Give them this form or have them make this format on a sheet of paper. Let them work a couple of minutes and then transfer all their functions to Post-it® Notes.

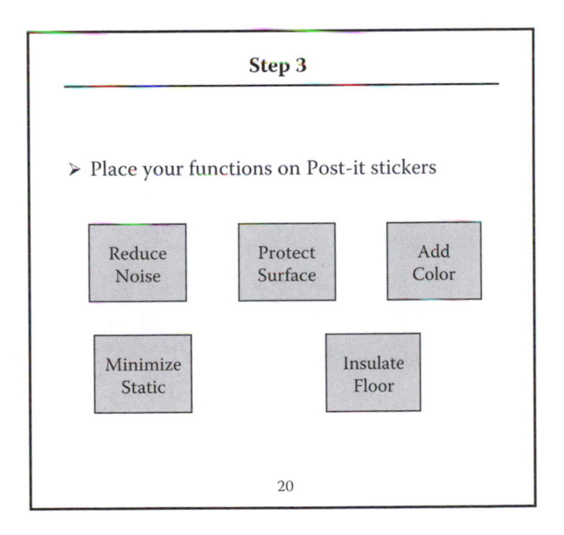


**Instructor notes:** Show these functions to the students. Have them add them to their list if they haven't already thought of them.

Hold back the next slide (21) until they work on their FAST diagram on the wall.

Have students organize their functions on stickers into a FAST diagram. The objective is to determine the Basic function for their carpet project.

You will find that "Things that happen all the Time" or "Project Criteria" will be things like:

- PROTECT SURFACE
- ADD COLOR
- INSULATE FLOOR

REDUCE NOISE does not happen all the time. It only happens when people are walking on the carpet! That should be the out-of-scope goal.

CUSHION FOOTSTEPS is how we are reducing noise. That becomes at least one of the Basic functions of the carpeting project.

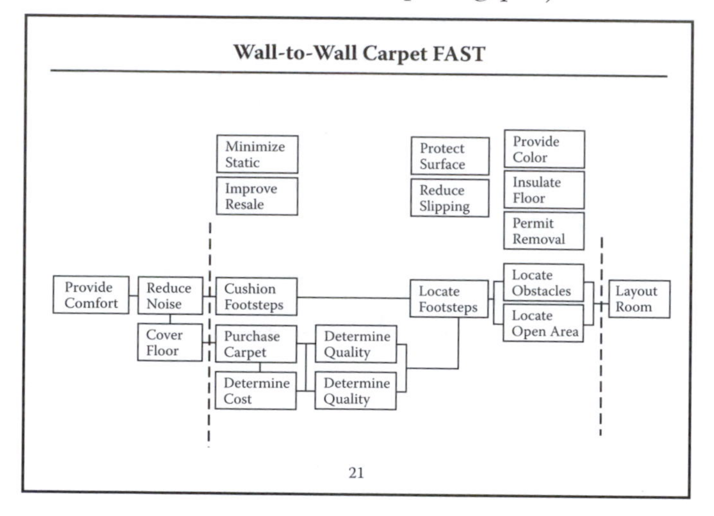


**Instructor notes:** Notice that none of the nouns that indicate method of achievement are on the critical path. The line below the critical path uses the terms (from left to right) *floor* and *carpet*. These indicate methods of achievement and are not purely unattached functions.

**The key question comes from this diagram:** "How does one cushion footsteps?" The answer is by LOCATING FOOTSTEPS. You can't cushion what you can't locate!

**Ask the students:** "Where in any room will you find no footsteps (or very few) being taken?"

Hold back the next slide (22) until you receive sufficient student response.

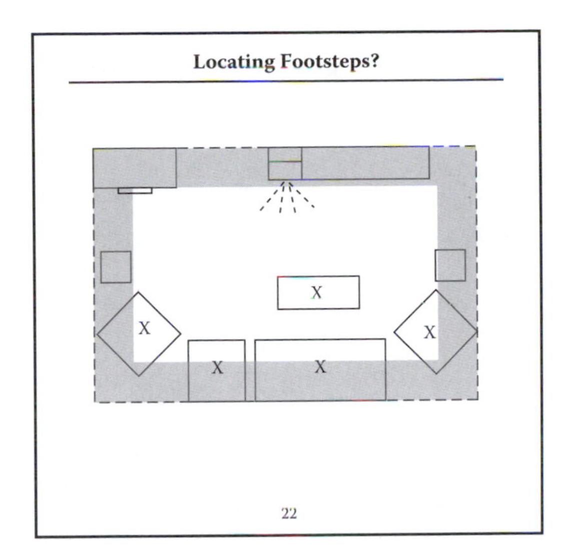


The shaded delineates the two-foot area along each wall. The furniture placement is outlined.

**Instructor notes:** Where don't footsteps ever occur in a room? Answer: under the furniture!

Is there anywhere else in a room that footsteps normally don't occur? Answer: around the edge of the wall!

People don't skinny up to the wall and walk right alongside it for its full length and then turn 90 degrees and continue walking.

So, how far away from the wall don't they walk? Answer: about 2 feet.

Well then, what is it worth to CUSHION FOOTSTEPS two feet away from every wall? Answer: nothing.

That, in effect, reduces a 10 × 15 room size (150 square feet) down to a 8 × 11 room size (88 square feet). A 58% reduction!

Give the students a blank Function Analysis Worksheet to fill out.

Hold back the next slide (23) until you see what they do with the worksheet.

**Step 4. Allocating Cost & Worth to Function**

| Item or Component | Verb | Noun | Kind | Function Cost | Tentative Alternate | Function Worth | Value Index |
|---|---|---|---|---|---|---|---|
| Wall-to-Wall Carpet | | | | | | | |
| Pad | Absorb | Noise | S | $ 96.83 | Eliminate | $ 1.00 | 96.8 |
| Carpet | Cushion | Footsteps | B | $ 608.33 | Issue Slippers | $ 20.00 | 30.4 |
| | Small | Order | S | $ 134.55 | Eliminate | $ 1.00 | 134.6 |
| | | | | $ 839.71 | | $ 22.00 | 38.2 |

LS BB Refine

23

**Instructor note:** This is an example of a filled out function analysis worksheet for this carpet example.

**Brainstorm the Function**

What are alternative ways to:

**Cushion Footsteps**

24

**Instructor note:** Have the students brainstorm this function. The actual solution was the purchase of a 6 × 9 carpet.

This reduced the area to CUSHION FOOTSTEPS from 150 square feet to 54 square feet—a 64% reduction.

**Homeowner's Solution**

**Purchase a 6 × 9 rug**

- Benefits:
  - Eliminates paid labor to install.
  - Eliminates cost of partial roll usage.
  - Reduces size of pad (if needed).
  - Retains beauty of hardwood surrounding floor.
  - If an oriental rug, it increases in value.
  - You can take it with you when you move again!

25

**Instructor note:** As you can see from this solution, there were substantial benefits other than just the cost savings.

When this implemented action is first presented, someone in the audience may say, "You didn't life-cycle cost this decision. You need two sets of cleaning equipment for maintaining both carpeting and hardwood now!"

The homeowner's response is: "Don't you think we already have two sets of cleaning equipment?" and "How much an hour do you think my wife gets for cleaning?"

Don't forget, the user—the customer—determines value and she knew a good deal when presented with it because she bought an oriental rug for the living room and cheaper area rugs for the rest of the house. Of course, those will eventually be replaced as the homeowner's revenue increases.

**Summary: Key Insights from Using VA**

- Value mismatches by focusing on cost and worth
- Tradeable requirements through function analysis and brainstorming
- Effective implementation approaches by classifying functions based on customer requirements

26

Value Analysis (VA) and Lean Six Sigma (LSS) develop solutions to problems from different perspectives, and therefore use of VA may provide some additional insights. Some of the most important distinctions are as follows:

- VA explicitly considers cost by collecting cost data and using cost models to make estimates for all functions over the life cycle. LSS reduces cost by eliminating waste and reducing variation through the use of statistical tools on process performance data. Exclusive emphasis on waste can be contradictory to reducing life-cycle cost. In VA, some waste can be tolerated if it is necessary to achieve a function that reduces the life-cycle cost. Safety stock to mitigate occasional supply disruption is a good example.
- In determining what should be changed, VA's function analysis identifies areas that cost more than they are worth, while LSS identifies root causes of problems or variations. VA's separation of function from implementation forces engineers to understand and deliver the requirements.
- For required functions that cost more than they are worth, VA uses structured brainstorming to determine alternative ways of performing them. LSS brainstorms to identify how to fix the root causes. Because functional thinking is not the common way of examining products or processes, VA augments the structured innovation process in a way that generates a large number of ideas. Enormous improvements are possible by determining which functions are really required and then determining how to best achieve them.
- VA develops solutions by evaluating the feasibility and effectiveness of the alternatives. LSS emphasizes solutions that eliminate waste and variation and sustain the achieved gains. VA eliminates waste in a different way. VA separates the costs required for basic function performance from those incurred for secondary functions to eliminate as many non-value-added secondary functions as possible, improve the value of the remaining ones, and still meet the customer requirements.
- An LSS focus on quick wins may preclude an in-depth analysis of the situation. Without analysis, projects can suboptimize or even work in opposition to one another. Using function analysis should prevent this suboptimization.

**Summary: Areas Benefiting from VA Insights**

- Producing a product
  - Conceptual decision and design
  - Preliminary design
  - Detailed design
  - Production
  - Operations
- Providing a service
  - Conceptual design
  - Operations
- Executing a construction project
  - Preliminary design
  - Detailed design

27

**Producing a product:** A concept decision determines an overarching approach to meet a capability need. By considering function and cost, a VA approach can provide important insights, and function analysis determines what must be done. VA links the customer requirements to the design to manage cost. Companies worldwide integrate VA concepts into their design processes to establish target costs and ensure that unnecessary functions and requirements are eliminated. Production costs can often be reduced by introducing new technologies, new processes, new materials, and/or new designs. In the operations and support phase of the product life cycle, VA provides additional opportunities to enhance options. VA concepts can identify a large number of resolution options, evaluate their potential for solving the problem, develop recommendations, and provide incentives for the investments needed for successful implementation.

**Providing a service:** VA application to the design or redesign of a service (and by analogy a process) is similar to the product situation. Using VA to challenge the requirements creates opportunities to improve upon other solutions. Function analysis challenges requirements by questioning the existing system, encouraging critical thinking, and developing innovative solutions. It ensures that areas of major expenditure receive attention in the early stages of a service contract.

**Executing a construction project:** VA has the greatest potential when the approach and costs are known (e.g., after design definition, approved feasibility study, and early remedial design are completed). VA identifies the essential functions and derives lower-cost alternative ways of accomplishing them. It brings the design team and client together to review the proposed design solutions, the cost estimate, and the proposed implementation schedule and approach, with the goal of achieving the best value for the money. The definition of what is good value on any particular project will change from client to client and project to project.

# Appendix E: Process-Oriented VE Training Material for LSS Black Belt Training

**Selecting a Process for Study**

Ask the following questions to determine if the selected study is a process and if it is appropriate for study:

1. **Is it a process?**
   - Does it currently exist in the organization?
   - Can it be flowcharted?

   If **No** to any question – it is not a process!

2. **Does the process recycle frequently?**
   - Is there less than one month between the first step and the last step?
   - Does the process recur several times in a day or week or every month?

   If **No** to any question – it is not an appropriate subject for study!

1

Value analysis may be used on a process as well as a product. Before proceeding, however, there are certain criteria to consider to determine if it is appropriate for study.

**Selecting a Process for Study**

3. **Can the process be measured easily?**
   - Is there a paper trail that follows the process?
   - Is there an observable beginning and end to the process which can be detected and measured?

   If **No** to any question – there will not be sufficient information or data for the study!
4. **Is the process currently undergoing transition?**
   - Has another group been assigned to work on the process?
   - Will another group be assigned to work on it?

   If **Yes** to any question – it is not appropriate for study at this time!

2

**Defining the Process**

**State, Identify, or List the Following for the Process:**

- **Output** of the process
- **Customer**(s) of the process output
- **Requirements** of the customer
- **Process Participants**
- **Process Owner**
- **Stake Holders**
- **Process Boundaries** (First step, Last step)
- **Inputs** and their **Suppliers**
- **Cycle Time** for the process

3

Once it is determined that the process is suitable for study, certain data on the process should be collected in order to construct a Function Analysis System Technique (FAST) diagram.

**Process Function – Cost-Worth**

➢ Value analysis steps

1. Determine process cost (Time)
2. Allocate cost to function
3. Determine worth of the functions
4. Calculate the value index

4

With a product, you normally will know its cost and price. However, with a process, its cost will not be as apparent. These costs are normally internal and represent the man-hours or man-days for the service being performed. Management knows how many people they have, but may not have broken down their work effort in detail to know how long each step of the process takes and what effort goes into it.

To do this you need a process flowchart so that you can do the following:

- Look at function costs (time)
- Identify value mismatches

A value mismatch is where it is apparent that cost does not equal worth.

Using this information to calculate the value index for each function should help to prioritize what might be most beneficial step in the process to work on to improve.

**Process Example**

Hospital Records Management System

Statistics:

- 643 Beds
- 37,770 new patient records/yr
- 253,092 old patient records/yr
- 512,256 patients/yr seen
- Of these
  - 56% are primary care
  - 32% are outpatients
  - 5% are inpatients
  - 7% are accident/emergency

5

Using this example, we will describe each of the steps in the previous slide. Our example is the records management system for a large metropolitan hospital that has the annual number of patients served as shown on this slide.

**Determine Process Cost**

For our hospital records management study a process cost estimate was developed.

This was approved by management before starting the study.

**Estimate**

| Departments | Labor Hours/year |
|---|---|
| Medical Records | 112,112 |
| Computer Services | 100,672 |
| Registration | 34,320 |
| Inpatient Admission | 11,440 |
| Primary Care | 45,760 |
| Accident & Emergency | 2,288 |
| Outpatient Admission | 50,336 |
| X-ray | 57,200 |
| Laboratory | 57,200 |
| Central Transcription | 22,880 |
| | 494,208 |

Other Costs
Medical Record Supplies
Contracted Services
Computer Supplies
Depreciation
Overhead

6

The Value Engineering (VE) team prepared this process cost estimate and found that more than 494,000 hours per year were spent on records management.

It is important that management (in this case, hospital management) accept and approve the estimate of their system before conducting the study so that they know the magnitude of the problem being studied and believe the impact of any recommendations resulting from the study.

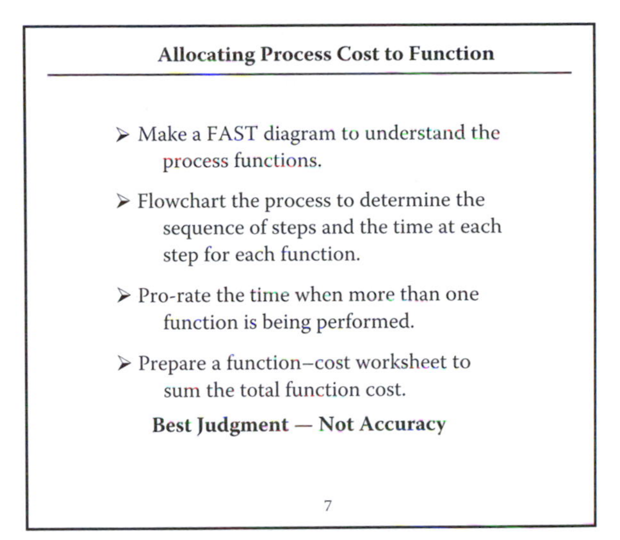


In value analyzing a process, process cost is time. And, when you determine the time to perform each function, you should be aware of the salary of the person doing it, the materials and tools they use to do their job, and the overhead of the organization for which they work.

Use best judgment to obtain a reasonable time allocation for the function performed. The best way to do this is to observe the process and interview each person in the process to determine what they are doing, why they think they are doing it, how long they take to perform the function, and what lag time they have before repeating the function they are performing.

Doing this on a rational basis is more important than explicit accuracy.

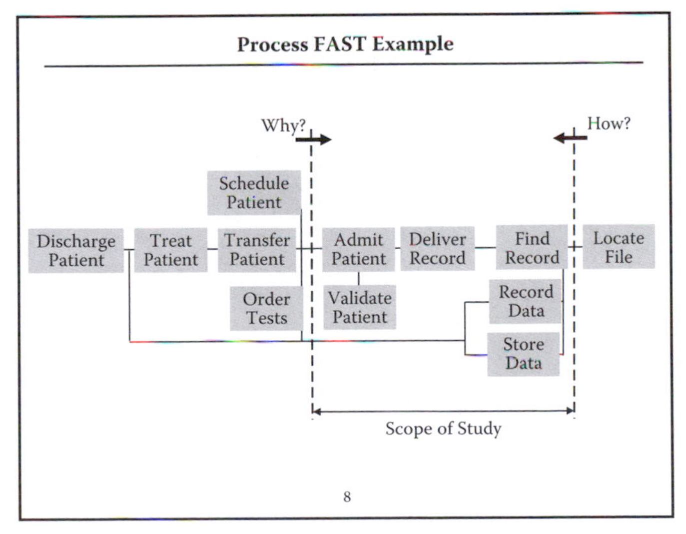


The scope of the value analysis study is shown by this FAST diagram.

The Why–How is to find the record for each patient returning to the hospital in order to validate it so the patient can be transferred from admissions to the proper office for treatment. They prepare a new record for each new patient for input into their system for validation.

This diagram was prepared by consolidating individual tasks into relevant functions that best describe the overall activity that is occurring. For example, many of the functions on this diagram can be expanded into much more detail; for example, FIND RECORD involves the SEARCHING function, OPENING FILES, TRANSFERRING RECORDS, etc.

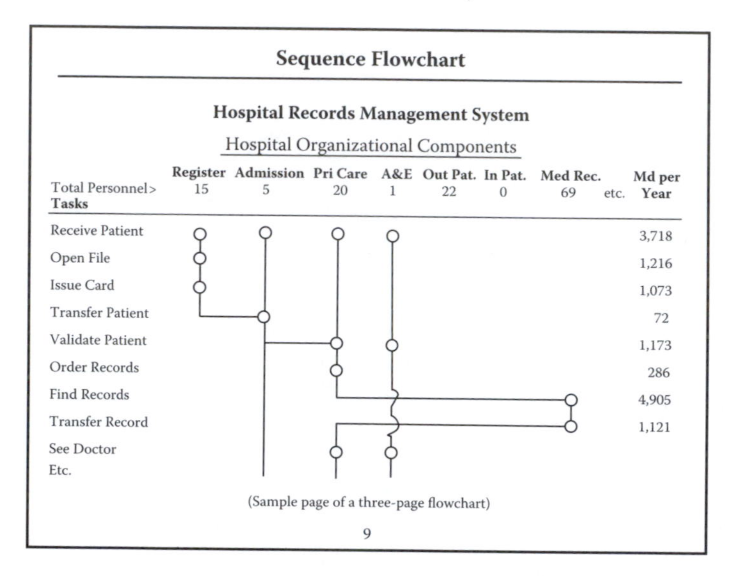


Here is the start of the three-page flowchart developed for the study.

It was developed by interviewing personnel from each department involved with patient record keeping. The tasks, as the employees described them, are detailed functions that were consolidated into more basic functions to describe the overall patient records operation as it was occurring.

**Function Cost**

**All Cost in Man-Days per Year**

| **Components** | **Total Cost (Md/Yr)** | **Admit Patient** | **Validate Patient** | **Transfer Patient** | **Schedule Patient** | **Find Record** | **Deliver Record** | **Order Tests** | **Dischge Patient** | **Record Data** | **Store Data** |
|---|---|---|---|---|---|---|---|---|---|---|---|
| Registration | 4,290 | 4,290 | | | | | | | | | |
| Admission | 1,430 | 270 | 144 | 144 | 144 | 500 | 72 | | 156 | | |
| Primary Care | 5,720 | | 1,144 | 572 | 1,430 | 286 | | 286 | | 2,002 | |
| Accident & Emergency | 286 | 214 | 29 | 43 | | | | | | | |
| Outpatient | 6,292 | | 629 | | 4,090 | | | 944 | | | 629 |
| Medical Records | 13,156 | | | | | 5,920 | 1,315 | | | 1,974 | 3,947 |
| X-Ray | 7,150 | | | | 2,860 | 2,145 | | | | 1,430 | 715 |
| Lab | 7,150 | | | | | | | 3,575 | | 3,575 | |
| Typing Pool | 2,860 | | | | | | | | | 2,860 | |
| Computer Department | 30,888 | | | | | | | | | | 30,888 |
| | 79,222 | 4,774 | 1,946 | 759 | 8,524 | 8,851 | 1,387 | 4,805 | 156 | 11,841 | 36,179 |
| Supervision | 5,425 | | | | | 283 | | | | 283 | 4,859 |
| | 84,647 | 4,774 | 1,946 | 759 | 8,524 | 9,134 | 1,387 | 4,805 | 156 | 12,124 | 41,038 |
| **Patients per Year** | | 124,200 | 512,200 | 28,700 | 475,600 | 475,600 | 951,200 | 1,000,000 | 188,700 | 512,200 | 475,600 |
| **Staff Minutes per Patient** | | 18.5 | 1.8 | 12.7 | 8.6 | 9.2 | 0.7 | 2.3 | 0.4 | 11.4 | 41.4 |

10

In a process, many functions are repeated by different people in different departments throughout the process.

This chart accumulates all the time spent for each function from the flowchart. Functions are listed across the top of the chart and departments (organizational components) down the left side.

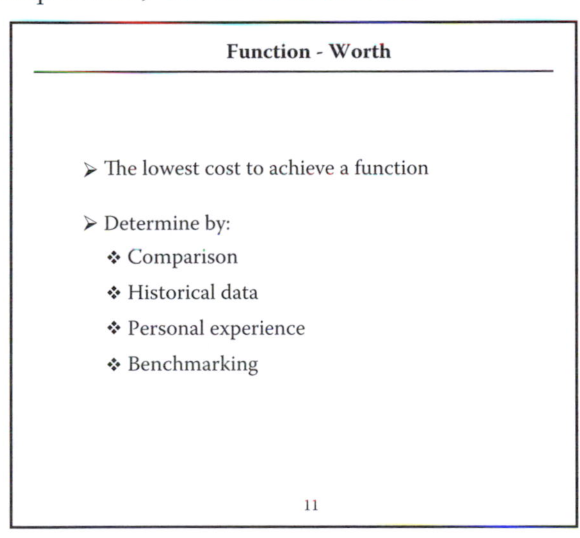


Worth is just a technique, not an absolute value.

- It is based upon the evaluator's judgment and experience.
- It is used only as a tool to identify the value index relationships of functions.

It would be presumptuous to attempt to tell someone the worth of a system. However, it is not presumptuous to tell them the worth of a function.

In a process, the fastest method that you can think of to achieve a necessary function is used to indicate the worth of that function.

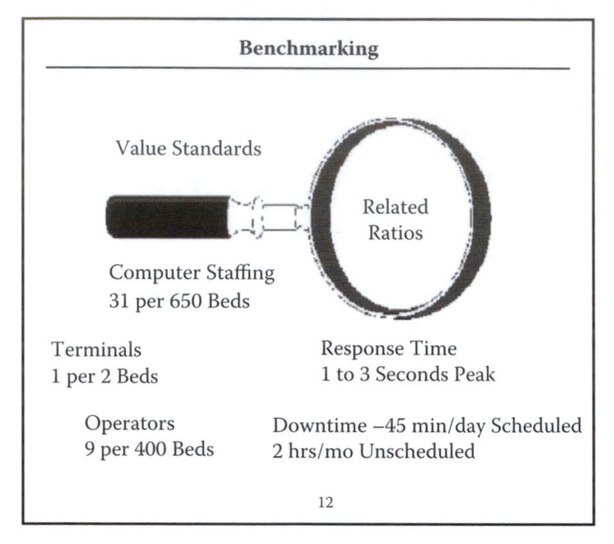


Value standards developed by experienced organizations can also be used as indicators of overall worth. These standards are also called *benchmarks* since they reveal a goal that others have achieved.

In our hospital patient records management study, an outside expert joined the study and provided the benchmark data achieved by other hospitals in their patient records management system. These data are useful to the team in judging the worth of the system being studied.

**Evaluate by Worth**

- Worth: The lowest cost (time) that it will take to perform the function
  - What else does that function? How long does that take?
  - What performs a function similar to it? What time does it take?
  - Worth evaluation is a quick approach to identifying functions that need value improvement.

13

Think of the fastest time that you know of that would serve the function.

Without regard to failure, that fastest time is a comparative worth of the function. If you feel that the function is not necessary for any reason, then the worth of the function is zero.

**Function Worth & Value Index**

| Components | Total Cost (Md/Yr) | Admit Patient | Validate Patient | Transfer Patient | Schedule Patient | Find Record | Deliver Record | Order Tests | Dischge Patient | Record Data | Store Data |
|---|---|---|---|---|---|---|---|---|---|---|---|
| Registration | 4,290 | 4,290 | | | | | | | | | |
| Admission | 1,430 | 270 | 144 | 144 | 144 | 500 | 72 | | 156 | | |
| Primary Care | 5,720 | | 1,144 | 572 | 1,430 | 286 | | 286 | | 2,002 | |
| Accident & Emergency | 286 | 214 | 29 | 43 | | | | | | | |
| Outpatient | 6,292 | | 629 | | 4,090 | | | 944 | | | 629 |
| Medical Records | 13,156 | | | | | 5,920 | 1,315 | | | 1,974 | 3,947 |
| X-Ray | 7,150 | | | | 2,860 | 2,145 | | | | 1,430 | 715 |
| Lab | 7,150 | | | | | | | 3,575 | | 3,575 | |
| Typing Pool | 2,860 | | | | | | | | | 2,860 | |
| Computer Department | 30,888 | | | | | | | | | | 30,888 |
| | 79,222 | 4,774 | 1,946 | 759 | 8,524 | 8,851 | 1,387 | 4,805 | 156 | 11,841 | 36,179 |
| Supervision | 5,425 | | | | | 283 | | | | 283 | 4,859 |
| | 84,647 | 4,774 | 1,946 | 759 | 8,524 | 9,134 | 1,387 | 4,805 | 156 | 12,124 | 41,038 |
| **Patients per Year** | | 124,200 | 512,200 | 28,700 | 475,600 | 475,600 | 951,200 | 1,000,000 | 188,700 | 512,200 | 475,600 |
| **Staff Minutes per Patient** | | 18.5 | 1.8 | 12.7 | 8.6 | 9.2 | 0.7 | 2.3 | 0.4 | 11.4 | 41.4 |
| **Worth (Md/Yr)** | 55,545 | 1,700 | 940 | 759 | 6,800 | 990 | 0 | 2,400 | 156 | 9,800 | 32,000 |
| **Value Index** | 1.52 | 2.81 | 2.07 | 1.00 | 1.25 | 9.23 | --- | 2.00 | 1.00 | 1.24 | 1.28 |

14

The next step in the study is to judge the worth of each function.

Notice that the function, DELIVER RECORD, is judged to be of zero worth. The team judged that records can be delivered by computer screen rather than physically and that this should not take 1,387 man days per year.

Notice that the poorest value occurred in FINDING RECORD—a record that the hospital already had. This had a value index of 9.23.

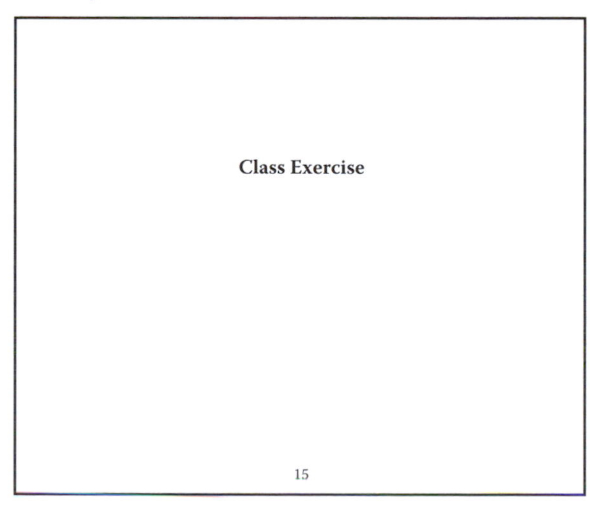

**Process Exercise**

**Packaging Process for Sheet Stamps**

1. Stamps are printed on large rolls of paper.
2. Rolls are delivered by fork truck to the packaging room.
3. Rolls are cut into sheets containing 400 stamps each.
4. Sheets are quartered containing 100 stamps each.
5. Quarters are shrink-wrapped and packed into cartons containing 250,000 stamps per carton.
6. Stamps are inspected twice, once after each cutting.
7. Stamps are inventoried twice, the rolls are counted coming in, and the cartons and rejected stamps are counted leaving the packaging room for shredding.

16

Here is the situation: You are doing a VE study for the Bureau of Printing and Engraving. They manufacture postage stamps. They have a labor-intensive process to package those stamps in cartons to send them out to post offices around the country.

Your assignment is not to change the stamps or the box. Just improve the efficiency of putting them into the box.

**Process Exercise**

**Packaging Process for Sheet Stamps**

| Production Staffing: | Other Cost: |
|---|---|
| • Distributors – 3 | • Supervision – 10% |
| • Examiners – 27 | • Materials – 27% |
| • Bookbinders – 2 | • Clerical – 1% |
| • Packers – 8 | • Depreciation –10% |
| • Stock Controllers – 4 | • Overhead – 8% |
| Staffing = 43% of cost | Other Cost Total = 57% |

17

Here is a list of the number of people involved in the packaging process and their jobs. Management accepted these labor costs as well as the identification of the other costs.

**Process Exercise**

**Function Cost in Manhours per Day**

**What Do You Think Is the Worth of Each Function?**

| Components | Total Cost Mh/Day | Load Sheets | Move Product | Uload Sheets | Inspect Sheets | Handle Defects | Count Product | Cut Sheets | Wrap Sheets | Box Product | Keep Record |
|---|---|---|---|---|---|---|---|---|---|---|---|
| Distributor | 24 | 6 | 7 | 11 | | | | | | | |
| Examiner | 216 | 6 | 51 | | 112 | 17 | 36 | | | | |
| Bookbinder | 16 | | 1 | 3 | | | | 12 | | | |
| Packer | 64 | | 14 | | | | | | 14 | 35 | 1 |
| Stock Controller | 32 | | 3 | | | | 12 | | | | 17 |
| Cost MH/Day = | 352 | 12 | 76 | 14 | 112 | 17 | 48 | 12 | 14 | 35 | 18 |
| **Worth Mh/Day =** | | | | | | | | | | | |
| **Value Index =** | | | | | | | | | | | |

18

The packaging operation works 24 hours per day. In a typical 24-hour day the packaging operation packaged 314 cartons of stamps.

Here is the cost (time) information for each function performed in the packaging operation. What do you think is the worth of each function and the value index for each function?

For example, what other ways are there to LOAD and UNLOAD SHEETS? How would a newspaper do it and would they spend 26 man hours per day doing it?

For example, would you spend 48 man hours per day to COUNT PRODUCT? Brainstorm what this might be worth by thinking how long it takes to COUNT PRODUCT if a machine were used.

**Function Worth Exercise**

| Function | Cost | Worth | Rationale for assignment of worth | VI |
|---|---|---|---|---|
| Load Sheets | 12 | 1 | Could be done automatically from cutter | 12.0 |
| Move Product | 76 | 44 | Use conveyor to move cut sheets from station to station | 1.7 |
| Unload Sheets | 14 | 1 | Could be done automatically from inspection station | 14.0 |
| Inspect Sheets | 112 | 75 | Don't handle each sheet, let them slide across viewing plate – doubles productivity | 1.5 |
| Handle Defects | 17 | 2 | Don't handle, let defective sheets be shuttled aside automatically | 8.5 |
| Count Product | 48 | 28 | Don't count rejected stamps – only good stamps that are boxed. | 1.7 |
| Cut Sheets | 12 | 11 | Slight efficiency improvement | 1.1 |
| Wrap Sheets | 14 | 14 | Good value | 1.0 |
| Box Product | 35 | 31 | Slight efficiency improvement | 1.1 |
| Keep Record | 18 | 9 | Log in rolls only and boxes out, 1 time per shift | 2.0 |

19

Worth is just a technique, not an absolute value.

- It is based upon the evaluator's judgment and experience.
- It is used only as a tool to identify the value-index relationships of functions.

It would be presumptuous to attempt to tell someone the worth of a system. However, it is not presumptuous to tell them the worth of a function.

In a process, the fastest method that you can think of to achieve a necessary function is used to indicate the worth of that function.

**Process Exercise**

**Function Worth–Answer**

| Components | Total Cost Mh/Day | Load Sheets | Move Product | Uload Sheets | Inspect Sheets | Handle Defects | Count Product | Cut Sheets | Wrap Sheets | Box Product | Keep Record |
|---|---|---|---|---|---|---|---|---|---|---|---|
| Distributor | 24 | 6 | 7 | 11 | | | | | | | |
| Examiner | 216 | 6 | 51 | | 112 | 17 | 36 | | | | |
| Bookbinder | 16 | | 1 | 3 | | | | 12 | | | |
| Packer | 64 | | 14 | | | | | | 14 | 35 | 1 |
| Stock Controller | 32 | | 3 | | | | 12 | | | | 17 |
| Cost Mh/Day = | 352 | 12 | 76 | 14 | 112 | 17 | 48 | 12 | 14 | 35 | 18 |
| **Worth (Mh/day)** | 216 | 1 | 44 | 1 | 75 | 2 | 28 | 11 | 14 | 31 | 9 |
| **Value Index** | 1.6 | 12.0 | 1.7 | 14.0 | 1.5 | 8.5 | 1.7 | 1.1 | 1.0 | 1.1 | 2.0 |

20

The largest value indices (poorest value) are the 12.0 and 14.0 it takes to LOAD and UNLOAD SHEETS. This indicates something that is possibly ripe for automation.

These two functions go hand in hand. We will explore these further for our example.

Notice also that the Stock Controller spends a lot of time COUNTING PRODUCT (VI = 1.7) and KEEPING RECORD (VI = 2.0). This indicates potential for 70–100% value improvement.

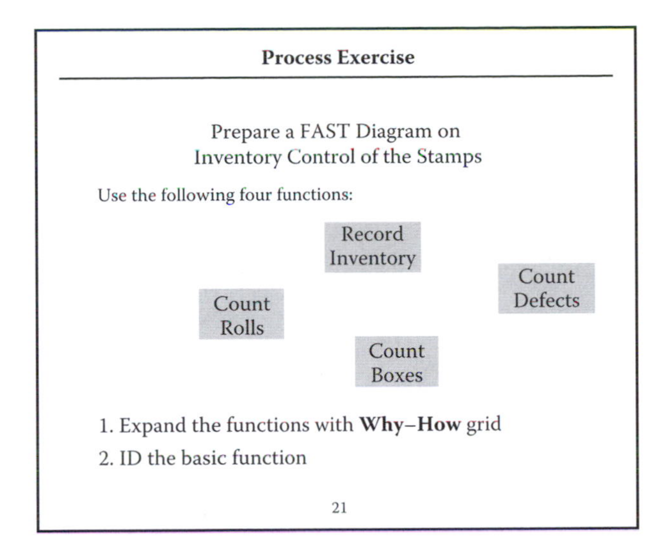


Let's make a FAST diagram of what the Stock Controller is doing. He is involved in the four functions that are displayed.

Expand each of these functions and identify the basic functions of the Stock Controller.

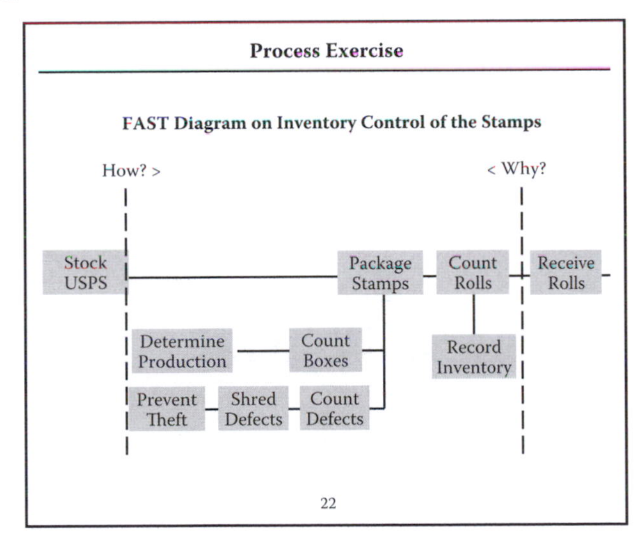


In its simplest form, here is a FAST diagram of what the Stock Controller is doing.

The function PACKAGE STAMPS can be expanded into another whole FAST diagram, but that is not necessary to see why the Stock Controller is COUNTING ROLLS, COUNTING BOXES, and COUNTING DEFECTS in order to ultimately DETERMINE PRODUCTION and PREVENT THEFT of individual defective stamps.

**Process Exercise**

- The functions **Keep Record** and **Count Product**
  - Cost 29 manhours per day
- The function **Prevent Theft** was the hidden reason that every stamp that left the room was counted for control purposes
  - **Yet**, only rolls, not stamps were counted when they came into the room

Brainstorm alternative ways to

**Prevent Theft**

of the defective stamps

23

Here are some ideas to PREVENT THEFT:

- Escort stamps leaving the room for the shredder.
- Inspect the pockets of all packaging workers before they leave the room.
- Install a camera system to prevent pilfering of individual stamps.

**Process Exercise**

**The Solution**

**Put a Shredder in the Packaging Room**

- Now, only count the rolls in and the number of cartons out
- Let the supervisor do this and give the stock controller other work to do on the crew

24

Sheets of stamps are rejected because of crimps in the sheet corners, the centering of perforations in the gutters is bad, the paper and gum seems bad, and sometimes a bit of ink is smudged during printing. Throwing away a sheet of 10 × 10 stamps (or 100 per sheet) will contain good individual stamps that could be salvaged.

Shredding the whole rejected sheet in the packaging room is a good idea because it reduces the temptation to take those stamps. It has the collateral benefit of reducing the need for an armed guard (not in the packaging budget) to escort the employee and those stamps when they leave the room for shredding.

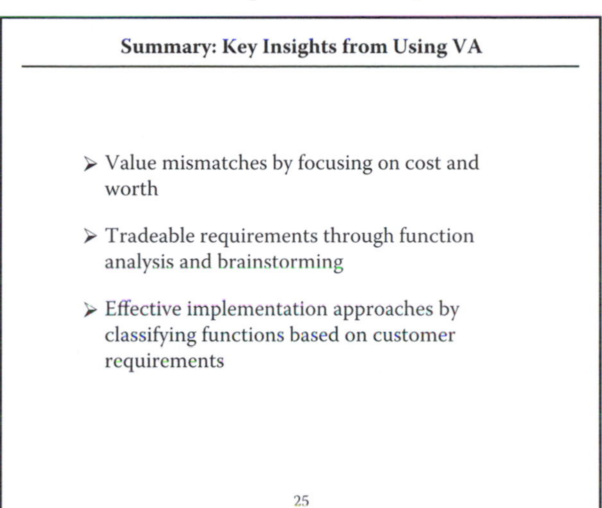


Value Analysis (VA) and Lean Six Sigma (LSS) develop solutions to problems from different perspectives, and therefore use of VA may provide some additional insights. Some of the most important distinctions are as follows:

- VA explicitly considers cost by collecting cost data and using cost models to make estimates for all functions over the life cycle. LSS reduces cost by eliminating waste and reducing variation through the use of statistical tools on process performance data. Exclusive emphasis on waste can be contradictory to reducing life-cycle cost. In VA, some waste can be tolerated if it is necessary to achieve a function that reduces the life-cycle cost. Safety stock to mitigate occasional supply disruption is a good example.
- In determining what should be changed, VA's function analysis identifies areas that cost more than they are worth, while LSS identifies root causes of problems or variations. VA's separation of function from implementation forces engineers to understand and deliver the requirements.
- For required functions that cost more than they are worth, VA uses structured brainstorming to determine alternative ways of performing them. LSS brainstorms to identify how to fix the root causes. Because functional thinking is not the common way of examining products or processes, VA augments the structured innovation process in a way that

generates a large number of ideas. Enormous improvements are possible by determining which functions are really required and then determining how to best achieve them.

- VA develops solutions by evaluating the feasibility and effectiveness of the alternatives. LSS emphasizes solutions that eliminate waste and variation and sustain the achieved gains. VA eliminates waste in a different way. VA separates the costs required for basic function performance from those incurred for secondary functions to eliminate as many non-value-added secondary functions as possible, improve the value of the remaining ones, and still meet the customer requirements.
- An LSS focus on quick wins may preclude an in-depth analysis of the situation. Without analysis, projects can suboptimize or even work in opposition to one another. Using function analysis should prevent this suboptimization.

**Summary: Areas Benefiting from VA Insights**

- **Producing a product**
  - Conceptual decision and design
  - Preliminary design
  - Detailed design
  - Production
  - Operations
- **Providing a service**
  - Conceptual design
  - Operations
- **Executing a construction project**
  - Preliminary design
  - Detailed design

26

**Producing a product:** A concept decision determines an overarching approach to meet a capability need. By considering function and cost, a VA approach can provide important insights, and function analysis determines what must be done. VA links the customer requirements to the design to manage cost. Companies worldwide integrate VA concepts into their design processes to establish target costs and ensure that unnecessary functions and requirements are eliminated. Production costs can often be reduced by introducing new technologies, new processes, new materials, or new designs. In the operations and support phase of the product life cycle, VA provides additional opportunities to enhance options. VA concepts can identify a large number of resolution options, evaluate their potential for solving the problem, develop recommendations, and provide incentives for the investments needed for successful implementation.

**Providing a service:** VA application to the design or redesign of a service (and by analogy a process) is similar to the product situation. Using VA to challenge the requirements creates opportunities to improve upon other solutions. Function analysis challenges requirements by questioning the existing system, encouraging critical thinking, and developing innovative solutions. It ensures that areas of major expenditure receive attention in the early stages of a service contract.

**Executing a construction project:** VA has the greatest potential when the approach and costs are known (e.g., after design definition, approved feasibility study, and early remedial design are completed). VA identifies the essential functions and derives lower-cost alternative ways of accomplishing them. It brings the design team and client together to review the proposed design solutions, the cost estimate, and the proposed implementation schedule and approach, with the goal of achieving the best value for the money. The definition of what is good value on any particular project will change from client to client and project to project.

# References


Ball, Henry A. June 2003. Value methodology—The link for modern management improvement tools. In *SAVE International 43rd Annual Conference Proceedings,* Scottsdale, Arizona, June 8–11, 2003.

Blocksom, Roland. 2004. STANDARD missile value engineering (VE) program—A best practices role model. *Defense AT&L Magazine*, July–August.

Bolton, James D., Don J. Gerhart, and Michael P. Holt. 2008. *Value methodology. A pocket guide to reduce cost and improve value through function analysis.* Lawrence, MA: GOAL/QPC.

Bozdogan, Kirkor. December 2003. *Lean aerospace initiative: A comparative review of lean thinking, Six Sigma, and related enterprise change models.* Center for Technology, Policy, and Industrial Development. Cambridge, MA: MIT.

Cell, Charles L., and Boris Arratia. 2003. *Creating value with lean thinking and value engineering.* Rock Island, IL: US Army Joint Munitions Command.

Chairman of the Joint Chiefs of Staff Instruction (CJCSI) 3170.01G. March 2009. *Joint capabilities integration and development system.*

Chang, Yuh-Huei, and Ching-Song Liou. June 2005. Implementing the risk analysis in evaluation phase to increase the project value. In *SAVE International 45th Annual Conference Proceedings*, San Diego, California, June 26–29, 2005.

Clarke, Dana W., Sr. June 1999. Integrating TRIZ with value engineering: Discovering alternative to traditional brainstorming and the selection and use of ideas. In *SAVE International 39th Annual Conference Proceedings,* San Antonio, Texas, June 27–30, 1999.

Cook, Michael J. June 2003. How to get Six Sigma companies to use VM and function analysis. In *SAVE International 43rd Annual Conference Proceedings,* Scottsdale, Arizona, June 8–11, 2003.

Crow, Kenneth. 2002. *Customer-focused development with QFD.* Palos Verdes, CA: DRM Associates.

Dull, C. Bernard. 1999. Comparing and combining value engineering and TRIZ techniques. In *SAVE International 39th Annual Conference Proceedings,* San Antonio, Texas, June 27–30, 1999.

Fraser, R. A. 1984. The value manager as change agent or how to be a good deviant. In *SAVE International Annual 24th Conference Proceedings,* Sacramento, California, May 6–9, 1984.

Gardner, Martha, and Gene Wiggs. 2007. Design for six sigma: The first 10 years. In Vol. 5 of *Proceedings of GT2007 ASME Turbo Expo 2007: Power for Land, Sea, and Air.* Montreal, Canada, May 14–17, 2007.

Goldratt, Eliyahu M. 1990. *Theory of constraints.* Croton-on-Hudson, NY: North River Press, Inc.

Hahn, Gerald J. 2008. *The role of statistics in business and industry.* Hoboken, NJ: John Wiley and Sons.

Hannan, Donald. May 2001. A hybrid approach to creativity. In *SAVE International 41st Annual Conference Proceedings,* Fort Lauderdale, Florida, May 6–9, 2001.

Hinckley, C. Martin. June 1998. *Managing product complexity: It's just a matter of time.* Report No. SAND-98-8564C. Livermore, CA: Sandia National Laboratories.

Hinckley, C. Martin. 2001. *Make no mistake.* New York: Productivity Press.

Hunt, Robert A., and Fernando B. Xavier. 2003. The leading edge in strategic QFD. *International Journal of Quality & Reliability Management* 20 (1): 56–73.

Huthwaite, Bart. 2004. *The Lean design solution.* Mackinac Island, MI: Institute for Lean Design.

Ishii, K., and S. Kmenta. (n.d.). *Life-cycle cost drivers and functional worth.* Project Report for ME317: Design for Manufacturing, Department of Mechanical Engineering. Palo Alto, CA: Stanford University.

Johnson, Gordon S. 2003. Conflicting or complementing? A comprehensive comparison of Six Sigma and value methodologies. In *SAVE International 43rd Annual Conference Proceedings,* Scottsdale, Arizona, June 8–11, 2003.

Kaufman, J. Jerry. 1990. *Value engineering for the practitioner.* Raleigh, NC: North Carolina State University.

Kaufman, J. Jerry, and James D. McCuish. 2002. Getting better solutions with brainstorming. In *SAVE International 42nd Annual Conference Proceedings,* Denver, Colorado, May 5–8, 2002.

Kolb, David A., and Richard E. Boyatzis. 1979. Goal setting and self-directed behavior change. In *Organizational psychology: A book of readings*, ed. David A. Kolb, Irwin M. Rubin, James M. McIntyre. Englewood Cliffs, NJ: Prentice-Hall.

Langley, Monica. 2006. Inside Mulally's "war room": A radical overhaul of Ford. *Wall Street Journal*, December 22.

Lehman, Theresa, and Paul Reiser. 2004. Maximizing value and minimizing waste: Value engineering and lean construction. In *SAVE International 44th Annual Conference Proceedings,* Montreal, Quebec, July 12–15, 2004.

Maass, Eric, and Patricia D. McNair. 2009. *Applying design for Six Sigma to software and hardware systems.* Upper Saddle River, NJ: Prentice Hall.

Mandelbaum, Jay, Royce R. Kneece, and Danny L. Reed. September 2008. *A partnership between value engineering and the diminishing manufacturing sources and material shortages community to reduce ownership costs.* IDA Document D-3598. Alexandria, VA: Institute for Defense Analyses.

Mandelbaum, Jay, Ina R. Merson, Danny L. Reed, James R. Vickers, and Lance M. Roark. June 2009. *Value engineering and service contracts.* IDA Document D-3733. Alexandria, VA: Institute for Defense Analyses.

Mandelbaum, Jay, and Heather Williams. 2010. "Synergy in enterprise change models: Opportunities for collaboration between Value Engineering and Lean/Six Sigma," http://www.value-eng.org/knowledge_bank/attachments/Mandelbaum%20&%20Williams%20-%20Synergy%20in%20Enterprise%20Change%20Models.pdf (accessed January 23, 2012).

Mudge, Arthur E. 1989. *Value engineering: A systematic approach.* Pittsburgh, PA: J. Pohl Associates.

Murman, Earl M. et al. 2002. *Lean enterprise value: Insights from MIT's lean aerospace initiative.* Houndmills, Basingstoke, Hampshire, UK: Palgrave.

Nomura Enterprise, Inc., and J. J. Kaufman Associates, Inc. (n.d.). *Function Analysis System Technique (FAST) student guide*. Prepared for the US Army Industrial Engineering Activity, Rock Island, Illinois.

Office of Management and Budget (OMB). 1992. Guidelines and discount rates for benefit-cost analysis of federal programs. OMB Circular A-94.

Parker, Donald E. 1998. *Value engineering theory*, rev. ed. Washington, DC: The Lawrence D. Miles Value Foundation.

Pucetas, John D. 1998. Keys to successful VE implementation. In *SAVE International 38th Annual Conference Proceedings*, Washington, DC: June, 14–17, 1998.

Reagan, Lisa A., and Mark J. Kiemele. 2008. *Design for Six Sigma—The tool guide for practitioners.* Bainbridge Island, WA: CTQ Media.

SAVE International. October 1998. *Function: Definition and analysis*, http://www.value-eng.org/pdf_docs/monographs/funcmono.pdf (accessed January 23, 2012).

SAVE International. 1999. *Function relationships: An overview.* SAVE International Monograph, http://www.value-eng.org/pdf_docs/monographs/funcrelat.pdf (accessed January 23, 2012).

SAVE International. June 2007. *Value standard and body of knowledge.* SAVE International Standard, http://www.scribd.com/doc/15563084/Value-Standard-and-Body-of-Knowledge (accessed January 23, 2012).

SAVE International. (n.d.). *Functional analysis systems techniques—The basics*, http://www.value-eng.org/pdf_docs/monographs/FAbasics.pdf (accessed January 23, 2012).

SAVE International. (n.d.). *Function logic models*, http://www.value-eng.org/pdf_docs/monographs/funclogic.pdf (accessed January 23, 2012).

Shingo, Shigeo. 1981. *Study of the Toyota production system from industrial engineering viewpoint.* Tokyo, Japan: Japanese Management Association.

Sicilia, J. D. (n.d.) "Champion Training" (briefing). Washington DC: Office of the Secretary of Defense.

Sloggy, John E. 2008. The value methodology: A critical short-term innovation strategy that drives long-term performance. In *SAVE International 48th Annual Conference Proceedings,* Reno, Nevada, June 9–12, 2008.

Snodgrass, Thomas J. 1993. Function analysis and quality management. In *SAVE International 33rd Annual Conference Proceedings*, Fort Lauderdale, Florida, May 2–5, 1993.

Sperling, Roger B. 1999. Enhancing creativity with pencil and paper. In *SAVE International 39th Annual Conference Proceedings,* San Antonio, Texas, June 27–30, 1999.

Stewart, Robert B. 2005. *Fundamentals of value methodology.* Bloomington, IN: Xlibris Corporation.

Thiry, Michel. 1997. *Value management practice.* Newtown Square, PA: Project Management Institute.

US Army. 1986. *Value engineering program management guide.* US Army Materiel Command Pamphlet 11-3.

US Department of Defense, Office of the Assistant Secretary of Defense. March 1986. *Value engineering.* DOD Handbook 4245.8-H.

US Department of Defense, December 2008. *Operation of the defense acquisition system.* DOD Instruction (DODI) 5000.02.

US Department of Defense. DOD Lean Six Sigma green belt course and the DOD Lean Six Sigma black belt course as contained in the training page of https://www.us.army.mil/suite/page/596053 (accessed January 23, 2012).

Vickers, James R., and Karen J. Gawron. 2009. A systems engineering approach to value engineering change proposals. Paper presented to the 2009 DMSMS Standardization Conference, Orlando, FL, September 21–24, 2009.

Wiggs, Gene. September 2005. Design for Six Sigma introduction. Paper presented to General Electric (GE) Aviation.

Woller, Jill Ann. 2005. Value analysis: An effective tool for organizational change. In *SAVE International 45th Annual Conference Proceedings,* San Diego, California, June 26–29, 2005.

Womack, James P., Daniel T. Jones, and Daniel Roos. 1990. *The machine that changed the world.* New York: Rawson Associates.

# Index

Page numbers in *italics* are for figures; page numbers in **bold** are for tables; page numbers followed by "n" are for footnotes.

**D**

**E**



**F**

**G**



**H**

## I



## J



## K



## L

## W



## X

# About the Authors

**Dr. Jay Mandelbaum** is a research staff member at the Institute for Defense Analyses (IDA). His research is focused on identifying best practices for value engineering in the Department of Defense (DOD), including providing recommendations for policy, guidance, and program management. He is also involved in research related to conducting technology readiness assessments in the context of the DOD acquisition process; establishing best quality assurance, manufacturing, and systems engineering practices for DOD programs; and improving how the DOD acquires new equipment and services.

Dr. Mandelbaum joined IDA in April 2004 after a 30-year career with the Federal Government. From 2002 to 2004, he was a member of the Systems Engineering staff in the Office of the Under Secretary of Defense for Acquisition, Technology and Logistics. He led the DOD's value engineering program and also managed an effort to reduce total ownership costs of defense systems.

Dr. Mandelbaum received his MS and DSc in operations research from The George Washington University in 1976 and 1982 respectively. He earned a BS in physics from Rutgers University in 1969.

**Anthony C. Hermes** is a research staff member at the Institute for Defense Analyses (IDA). He has eight years of work experience with the IDA, including support to the Department of Defense, the Department of Homeland Security, and the Department of Energy. He is currently working value-engineering issues for the Office of the Under Secretary of Defense for Acquisition, Technology, and Logistics.

Prior to coming to his current position, Mr. Hermes spent 26 years in the US Army where he held leadership and management positions. His experiences included major construction management; facilities engineering management; acquisition cost analysis; planning, programming, budgeting, and execution analysis; operations research analysis; and joint, inter-agency, and combined/coalition operations. Much of his value engineering experiences were gained while serving with the US Army Corps of Engineers, including as Chief, Base Realignment and Closure Office, Headquarters Corps of Engineers.

Mr. Hermes received a BS in engineering technology in 1970 and an MBA in operations research in 1980 from Texas A&M University.

**Donald E. Parker** is an independent building consultant performing due diligence services for prospective buyers of commercial property, programming and budgetary work for developers, and value engineering and estimating services for owners and designers.

In the last decade, Mr. Parker served as executive vice president of National Government Properties with responsibility for Property Management of its $280-million portfolio of leased office, warehouse, and clinic space. He was also an employee of the Federal Government for nearly 30 years. He served as Special Assistant to the Commissioner, Public Buildings Service at the General Services Administration, and directed their value engineering program and established value engineering in the design and construction industry. He is also author of the four value-engineering textbooks.

Mr. Parker received his BS in civil engineering from Northwestern University in 1960. He is a professional engineer, a certified cost engineer, and a certified value specialist (life).

**Heather W. Williams** is a research associate at the Institute for Defense Analyses where her research focus includes systems engineering and nuclear weapons policy.

Ms. Williams earned her BA in international relations and Russian studies from Boston University in 2004 and her MA in security policy studies from The George Washington University in 2008. She is currently a PhD candidate in the War Studies Department at King's College London and her thesis topic is "The Legacy of 'Trust but Verify' in US-Russia Arms Control."